GRASSFIRES

Fuel, weather and fire behaviour

SECOND EDITION

Phil Cheney and Andrew Sullivan

CSIRO
PUBLISHING

© CSIRO 2008

All rights reserved. Except under the conditions described in the *Australian Copyright Act* 1968 and subsequent amendments, no part of this publication may be reproduced, stored in a retrieval system or transmitted in any form or by any means, electronic, mechanical, photocopying, recording, duplicating or otherwise, without the prior permission of the copyright owner. Contact **CSIRO** PUBLISHING for all permission requests.

National Library of Australia Cataloguing-in-Publication entry

> Cheney, Phil, 1940–
> Grassfires : fuel, weather and fire behaviour / Phil Cheney, Andrew Sullivan.
>
> 2nd ed.
>
> 9780643093836 (pbk.)
>
> Includes index.
> Bibliography.
>
> Grassland fires – Australia.
> Grassland fires – Australia – Forecasting.
> Fire weather – Australia.
>
> Sullivan, Andrew, 1968–
>
> 632.180994

Published by and available from
CSIRO PUBLISHING
150 Oxford Street (PO Box 1139)
Collingwood VIC 3066
Australia

Telephone: +61 3 9662 7666
Local call: 1300 788 000 (Australia only)
Fax: +61 3 9662 7555
Email: publishing.sales@csiro.au
Website: www.publish.csiro.au

Front cover image: Annaburroo experimental fire C113, 21/8/1986, NT.
Back cover images (from left to right): Annaburroo experimental fire B184, 15/8/1986, NT; Aerial view of Crase fire, ACT, 23/1/1975; Suppression on racecourse fire, ACT, 9/1/1965.
All cover images © CSIRO.

Set in 11/13.5 Adobe Minion
Edited by Adrienne de Kretser
Cover and text design by James Kelly
Illustrations by Angela Halpin, Tibor Hegedis and Andrew Sullivan
Typeset by Desktop Concepts Pty Ltd, Melbourne
Printed in China by 1010 Printing International Ltd

The paper this book is printed on is certified by the Forest Stewardship Council (FSC) © 1996 FSC A.C. The FSC promotes environmentally responsible, socially beneficial and economically viable management of the world's forests.

Foreword

In Australia the term 'bushfire' is commonly used to describe fires burning in the landscape. Bushfires are classified as forest, scrub or grass fires, each with their own characteristics depending on the type of vegetation involved. Grassfires, exposed to the full force of the wind, sometimes cover huge areas if unimpeded, causing immense damage to stock, forage and fences. Bushfires have been the subject of reports and studies for at least 200 years. Although we have a good deal of general information about the major fires of the 20th century, the damage they caused and the weather and fuel conditions under which they burnt, we know far less about fires of lesser intensity which barely rated a mention in the press but have burnt over huge areas in parts of Australia and should form an important part of our fire history.

During the 20th century there were many organisational changes, starting shortly after Federation with the formation of bushfire brigades when farmers, graziers and local councils realised that collective rather than individual fire control efforts were essential. Embryo forest services also found that their protection activities warranted a closer study of fire behaviour under conditions ranging from those appropriate to the safe burning of firebreaks to the weather associated with destructive fires.

Although the dates of commencement vary from one region to another, the first half of the 1900s saw the formation, under various names, of bushfire committees charged with the co-ordination of effort in each state and territory. It is not possible to say for certain when science was first called upon to help those who bore the brunt of fireline activities. In the federal sphere, at least since the 1950s, fire research activities have been carried out by the Bureau of Meteorology, the Forestry and Timber Bureau, several divisions of CSIRO and state forestry agencies. Although many other organisations have contributed to knowledge of fire behaviour, the late Alan McArthur is acknowledged as the pioneer of this field. Since his death in 1978 there have been several major fire situations, including those of Ash Wednesday 1983, when extreme weather conditions have resulted in the level of fire danger exceeding that considered previously to represent the probable maximum.

It is certain that Alan McArthur would have welcomed the advances in understanding fire behaviour made by Phil Cheney and Andrew Sullivan with the help of their colleagues. While work is still proceeding to improve our knowledge of forest fire behaviour, this book offers the opportunity to incorporate into a general discussion the results of recent research into grassfires and their behaviour, and the changes to systems to predict grassfire spread and fire danger.

This work has been written for all throughout Australia who are concerned with fire in grasslands. The information includes advice for those who have not been through a major fire but need to prepare for the experience. It ranges from the discussion of the steps to be taken when investigating major fires to the means for deciding the nature and placement of firebreaks in grassland areas. There is something in this work for all who are concerned with the problems of bushfire protection, whether through individual study or, perhaps more realistically, through group discussion.

Harry Luke (1909–2000)
Sydney
September 1997

Contents

Foreword   *iii*
Preface to the 1st edition   *vii*
Preface to the 2nd edition   *ix*
Acknowledgements   *x*

1 Introduction   1
     Grassfires in Australia   *1*
     A historical view   *2*
     A more recent view   *4*
     Predicting grassfire spread   *4*

2 Fuel   7
     Tropical grasslands   *7*
     Tussock grasslands   *9*
     Hummock grasslands   *10*
     Improved pastures   *12*
     Crop lands   *14*
     Comparison of grazed and ungrazed pastures   *14*
     Grassland life cycles   *14*

3 Combustion of grassy fuels   17
     The combustion process   *17*
     Diffusion flames   *19*
     Heat yield from grassy fuel   *21*
     Moving grassfires   *21*
     Fire types   *22*
     Flame zone characteristics   *24*
     Fireline intensity   *29*

4 Fire behaviour   31
     Ignition   *31*
     Fire growth   *32*
     Fire shape   *35*
     Fuel characteristics   *35*
     Fuel moisture content   *38*
     Wind speed   *40*
     Slope   *43*
     Fire behaviour in spinifex   *44*

5	**Predicting fire spread**	**49**
	Fuel condition	*50*
	Grass curing state	*52*
	Dead fuel moisture content	*55*
	Wind speed	*57*
	Slope	*59*
	CSIRO grassland fire spread prediction system	*60*
6	**Local variation and erratic fire behaviour**	**67**
	Wind	*67*
	Interaction of fuel moisture with relative humidity	*78*
	Other factors that influence fire behaviour	*80*
7	**Fire danger**	**83**
	Fire spread and fire danger prediction	*85*
8	**Wildfires and their suppression**	**87**
	Past wildfire events	*88*
	Suppression of grassfires under extreme conditions	*96*
	Firebreaks	*102*
9	**Grassfire investigation**	**105**
	Wind direction	*106*
	Shape of the fire perimeter	*108*
	Determining the origin	*112*
	Power line ignitions	*112*
10	**Safety: myths, facts and fallacies**	**115**
	The fire consumed all the oxygen around me and I could not breathe	*116*
	Boiled alive	*119*
	The fire creates its own wind	*120*
	The fire was so hot it melted the engine block	*122*
	The green grass burns faster	*123*
	Exploding petrol tanks	*124*
	Great balls of fire	*127*
	Faster than a speeding bullet	*128*
	The fire was equivalent to eight Hiroshimas	*129*
	It can't happen here	*130*
	The red steer	*131*
	Glossary	*134*
	Bibliography and further reading	*142*
	Index	*147*

Preface to the 1st edition

This book has been written to present our current knowledge of the spread and behaviour of grassfires in a way that is attractive and informative to rural land holders and members of rural bushfire brigades. The book includes new information that has been obtained from 15 years of research into the behaviour and spread of grassfires by the Bushfire Behaviour and Management Group of CSIRO Forestry and Forest Products and recent work by other divisions of CSIRO and the Department of Conservation and Land Management (CALM) in Western Australia. Most of this work has been published in the scientific literature but there has been no comprehensive account of grassfires since Alan McArthur published *Weather and Grassland Fire Behaviour* in 1966.

In 1985 Australian rural fire authorities sponsored the Bushfire Behaviour and Management Group to undertake a large field investigation into grassland fire behaviour to resolve some of the problems they were having with the predictions of fire spread using the McArthur Grassland Fire Danger meters. The research was conducted in the Northern Territory and much of the information and many of the illustrations used in this book come from those experiments.

The research resulted in a major finding that the speed of a fire was influenced not only by weather and fuel variables but also by the size and shape of the fire itself, and demonstrated the need for scaling factors to be introduced into fire science. This issue had been largely ignored by fire scientists around the world and now explains many of the difficulties that arose from using the results of small-scale experiments to predict the behaviour of wildfires.

In this book we have described the important variables that influence grassland fire behaviour and how these are combined to predict grassfire spread both in the pastoral areas of southern Australia and in the open forests and woodlands of northern Australia. The section on the behaviour and spread of fire in spinifex grasslands has been drawn largely from the work of Neil Burrows and his colleagues in CALM and is included to illustrate the differences that arise when fire spreads through discontinuous grassy fuel.

We are indebted to many people who have assisted both in the experiments and in the writing of this book. We would like to thank the Bushfires Council of the Northern Territory, which provided very considerable field support for the experimental burning, including preparation of the experimental areas and control of individual fires; the Country Fire Authority, Victoria and the Country Fire Service, South Australia, which provided financial support for the field experiments; and particularly the staff of the Bushfire Behaviour and Management Group who assisted in the running of the experiments, the preparation of scientific papers and finally in the preparation of this book.

The preparation of this book was made possible by a grant from the SA Great CFS Training and Research Foundation. We hope that we have been able to bring a better understanding of grassland fire behaviour and aspects of wildfire control to a wide cross-section of the rural community in southern Australia and in all other areas where grass is the dominant fuel type.

Phil Cheney
Andrew Sullivan
CSIRO Forestry and Forest Products
Bushfire Behaviour and Management Group

Preface to the 2nd edition

It has been 11 years since the publication of the first edition of *Grassfires: Fuel, Weather and Fire Behaviour*. The book has been well received and used as training material and as a classroom text, then went out of print. Since its publication our work has focused on refining our understanding of the behaviour of forest fires. As a result, there is essentially no new material on the behaviour of grassfires.

However, the evolution of our understanding of fires in dry eucalypt forests has necessarily brought about a refinement of our understanding of the processes and mechanisms that influence the behaviour of bushfires in general, namely, combustion and the nature of the wind. Thus, when the opportunity arose to present a second edition of *Grassfires*, although there was no need to revise the basic fire behaviour information we did revise the discussion of the associated mechanics of grassland fire behaviour. We have also corrected a number of errors and general inadequacies in the first edition. We have added a new wildfire case history to bring the book up to date with some of the recent events in Australia and have expanded the discussion of the myths, facts and fallacies that surround grassfires to show how they may compromise fire safety.

Phil Cheney
Andrew Sullivan
CSIRO Sustainable Ecosystems
Bushfire Dynamics and Applications

Acknowledgements

We gratefully acknowledge the following for permission to reprint or adapt copyrighted material:

Primary Industry South Australia, Forestry: Figure 8.6 from 'Forest fires in South Australia on 16 February 1983 and consequent future forest management aims' by A. Keeves and D.R. Douglas in *Australian Forestry* **46**, 148–162 © 1983.

University of Adelaide, Department of Agronomy, Waite Institute: Figure 2.8 from Master of Agricultural Science Thesis 'The growth, senescence and ignitability of annual pastures' by R.T. Parrot © 1964.

Bureau of Meteorology: Figures 2.9 and 6.3 from 'Meteorological Aspects of the Ash Wednesday Fires 16 February 1983' by the Bureau of Meteorology © Commonwealth of Australia 1984, reproduced with permission.

International Association of Wildland Fire: Figures 3.2, 4.2 and 4.13 were first published in *International Journal of Wildland Fire* **12**(1), **5**(4) and **3**(1) respectively © IAWF 2003, 1995 and 1993.

1
INTRODUCTION

Grassfires in Australia

Fire is one of the most important elements in the Australian landscape. Many Australian plants have evolved with fire as a necessary part of their life cycles. It is estimated that, on average, an area of more than 2.5 million hectares is burnt each year by wildfires – and much larger areas in extreme years. The figures are somewhat uncertain, however, partly because it is difficult to distinguish between wildfires and agricultural/cultural burning-off in the tropics. In 1992, for example, 7.4 million ha was burnt in the Northern Territory – 5.5% of the total land area – but the fraction that was burnt by wildfires is unknown.

Most wildfires in Australia are called bushfires because they occur in rural areas or 'the bush'. It is more accurate, however, to describe a wildfire by using the dominant fuel type involved, e.g. forest fire, scrub fire or grassfire.

Grass is the most common fuel type in Australia. Nearly 75% of the country is grassland of one species or another, with half of that area used for sheep or cattle grazing. Grasslands range in type from the vast open Mitchell grass plains of northern Australia to the buttongrass moors of western Tasmania. They may be open treeless plains or areas that have been cleared for agriculture or grazing, and include improved pastures and crop lands of southern Australia. Grasses are also the dominant fuel in woodlands and open forests with a history of regular burning.

Figure 1.1: Progressive suppression and mop-up on the southern flank of a fire at Binalong, NSW, 18 December 1990.

During years of exceptionally heavy rainfall, grasses can form a continuous sward through the normally arid and semi-arid interior of the continent.

A historical view

Bushfires have been a major concern in Australia since Europeans arrived in the late 1700s. Repeated observations by early European explorers revealed that open forests and woodlands with a grassy understorey covered much of the land. They gained an impression of an annual conflagration during the fire season, caused by Aborigines. Historical accounts record that Aborigines used fire extensively for access, warfare, hunting, the promotion of edible plants, warmth, protection against snakes and insects and 'cleaning up the country'.

To the explorers, quite unused to burning on this scale, the Aborigines appeared to be careless with fire – they left camp fires unextinguished in hollows in large trees and stumps, dropped fire when travelling and so on. However, research into Aboriginal burning practices in the top end of the Northern Territory and in the Central Desert regions where there was still adequate memory of traditional practice has suggested that burning was done with considerable thought and great skill. Fire was applied in different ways. At times it was applied to specific areas under mild

Figure 1.2: Aerial photograph of Aboriginal burning in the Gibson Desert in 1953. Fifty per cent of the patches have an area of less than 5 ha. Many patches appear to have been lit from a line and have self-extinguished after running less than 2 km or have spread for less than 20 minutes at a critical wind speed enough to just maintain spread. Source: Royal Australian Air Force 1953.

conditions to ensure that only small patches burnt. At other times it was applied under more severe weather conditions and burning would continue for several days. Mostly it appeared to be applied to create a mosaic of relatively small patches of burnt and unburnt area (see Fig. 1.2). We believe that the Aborigines in these areas were acutely aware of the consequences of fire burning for months in extensive continuous dry grass, and covering huge areas. In desert areas this may destroy the habitat for small mammals for many years. Not only did they burn to create a suitable habitat for their food supply, but they also burnt to prevent the spread of large fires and to protect themselves. We do not have oral history of the burning practices in southern Australia but we consider that the same logic would prevail.

The arrival and spread of Europeans changed the population, distribution and food-gathering habits of the Aborigines and so caused a major change in burning regimes. The number and frequency of Aboriginal ignitions was drastically reduced, changing the fire regime from extensive annual or biennial patch-burning to one involving much longer periods between fires.

The reduced fire frequency allowed fuels to build up, so the intensity of fires increased. Deliberate firing of the landscape by farmers and graziers for clearing scrub and producing green pick for cattle and sheep occurred, but not frequently enough to mimic the fire regimes used by the Aborigines. Fires that occurred during mild weather in settled regions were quickly suppressed for fear of damage to assets such as homesteads, stock, crops and fencing. Those that broke out during extreme weather were more difficult to control. The absence of patch-burning by Aborigines meant that there were extensive areas of continuous fuel and fires often burnt for days, sometimes weeks, before a change of weather put them out.

A more recent view

Despite greatly improved equipment and organisation to fight fires, bushfires still cost Australia millions of dollars each year, with 72% of all bushfire damage costs the result of property damage due to grassfires.

Under extreme weather conditions, a grassfire can spread at more than 20 km/h, travel huge distances in a single day and threaten communities or other valuable assets many kilometres from its origin. For example, in South Australia in 1983 bushfires that started in open grassland travelled more than 65 km in four hours before burning into the extensive radiata pine plantations in the south-east of the state (see Fig. 8.6).

Grassfires can still burn huge areas of Australia, particularly through sparsely populated regions of low productivity. After the occasional very good growing season, grass may grow rapidly. Grazing has little impact on the fuel due to low stocking rates, and roads and firebreaks used to check fires in a normal season become quite ineffective. In 1974/75, after exceptionally heavy rainfalls during the preceding years, grassfires swept through the centre of the continent over a 6-month period and burnt a total of 117 million ha of countryside (Fig. 1.3).

Predicting grassfire spread

A number of fire danger rating systems have been developed to help warn the public of dangerous fire weather. In 1960 A.G. McArthur produced a set of grassland fire behaviour tables based on measurement of experimental fires. Fire danger classes described the difficulty of suppressing fires in average fuel, and the fire danger scale was linked directly to the rate of spread of the head fire. The McArthur grassland fire danger rating system, either in its original form or as modified by other workers, has been used since 1965 to give a quantitative index of grassland fire danger and provide public warnings. This system has been satisfactory for rating grassland fire

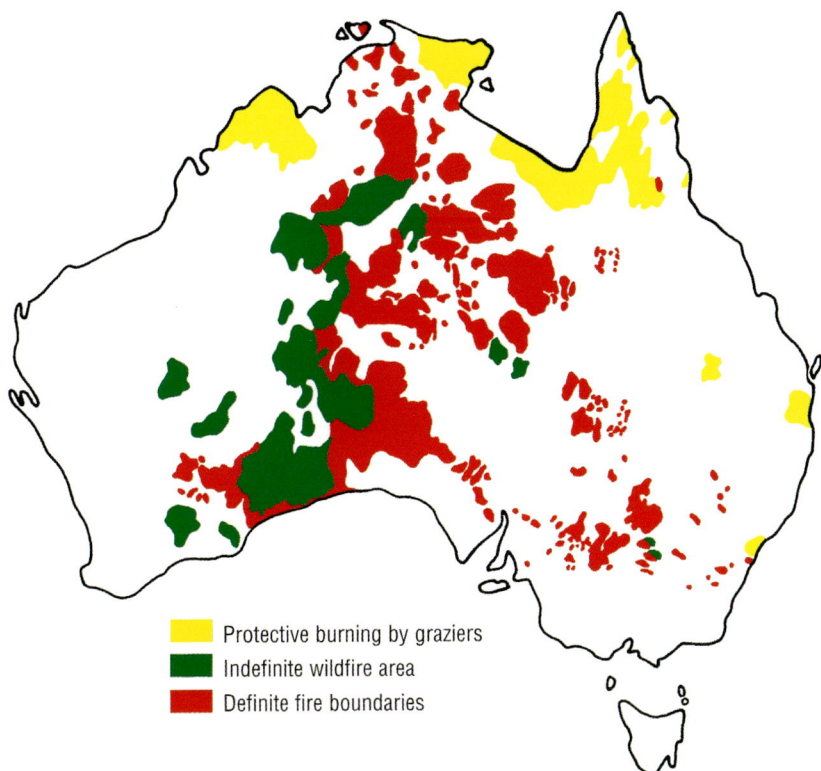

Figure 1.3: Area burnt in Australia during 1974/75 fire season. Source: Luke and McArthur (1978).

danger, but neither the Mk 4 nor the Mk 5 grassland fire danger meter gave satisfactory predictions of rate of spread. The introduction of fuel load in the Mk 5 meter led to better results for pastures that had been very heavily grazed. However, the Mk 5 meter did not perform as well as Mk 4 for light but continuous pastures. The fact that the two meters calculated different fire danger indices for the same weather conditions concerned the authorities responsible for issuing public warnings of fire danger. In 1986 rural fire authorities asked CSIRO to undertake a major study of grassland fire behaviour to resolve the problems in the existing systems and to design an improved fire spread prediction system that could be used in a wide range of grassy fuel types.

Although grass fuels are relatively simple compared with forest and scrubland fuels, different species of grasses form a wide range of structural types which generate different fire behaviour characteristics. The characteristics of grass fuels make them highly responsive to changes in weather variables – in particular wind speed, wind direction and relative humidity – and this often gives firefighters the impression that grassfires are highly variable and unpredictable. However, provided weather variables are known or can be predicted accurately for a given location, the general characteristics of grassfire behaviour – such as the average rate of spread, the fire's shape, the height of the flames and the distance the fire will travel – can be reasonably well predicted.

If firefighters are to successfully use a fire spread prediction system to make reasonably accurate forecasts of the behaviour and spread of grassfires, they need a sound understanding of the factors that influence grassfire behaviour and how these can vary across the landscape. This will enable them to appreciate the influence of factors not included in the system, and understand why observed rates of spread may vary from predicted values.

2
FUEL

Fire behaviour is determined by the physical structure of the fuel bed. Although many hundreds of grass species grow in Australia, for the purpose of predicting fire spread, grasslands can be classified into five broad groups with similar structural characteristics. These are:

- tropical grasslands;
- tussock grasslands;
- hummock grasslands;
- improved pastures;
- crop lands.

The main characteristic that influences the spread of fire in these groups is the continuity of the fuel bed, i.e. how unbroken the fuel bed is. The height of the grass has the greatest influence on flame height, and fuel load is the main factor affecting fire intensity. All these characteristics can be important in determining how difficult it will be to suppress a fire and its responsiveness to changes in the weather.

Tropical grasslands
These occur in high rainfall areas (greater than 750 mm per annum), and are mostly associated with open forests and woodlands characterised by numerous eucalypt

Figure 2.1: Sorghum intrans over 2 m high after flowering. These grasses collapse after curing, forming a fuel bed around 0.5–1.0 m high. The grass is mechanically flattened to reduce flame heights.

species, of which Darwin stringybark (*Eucalyptus tetrodonta*), Darwin woollybutt (*E. miniata*) and northern bloodwoods (e.g. *E. polycarpa*) are most typical.

The most common grass species are tall annual sorghums (*Sorghum intrans, S. stipodium*), but tall perennial grasses such as giant spear grass (*Heteropogon triticeus*) and perennial sorghum (*S. plumosum*) dominate in some areas. Numerous short grasses, both annual and perennial, occur but are generally sparse in distribution, especially where the taller grasses are dense.

The grass cover commonly grows to 3 m, and sometimes as high as 4 m, during the wet season. Grasses progressively cure with the onset of the dry season and collapse with the last rains to form a uniform fuel bed around 0.5–1.0 m high.

If not burnt, the annual grasses decompose almost completely during the next wet season and only a thin layer of organic material remains on the soil surface at the start of the following dry season. In open forests and woodlands, grass fuels are supplemented by scattered leaf litter and this can allow fires to burn early in the dry season when the annual grasses are only partially cured. The additional fuel can fill bare patches in the grass sward, making a more continuous fuel bed. Leaf fall during the dry season will also increase the total amount of fuel, thus increasing the intensity of late-season fires.

Grassfires in these fuels are generally extensive and continue burning overnight late in the dry season. Commonly more than 50% of the land area of the northern

Figure 2.2: A uniform sward of kangaroo grass in the Northern Territory. The top height of the pasture is around 1 m but the bulk of the grass is less than 30 cm.

regions of northern Australia, such as the Top End of the Northern Territory, is burnt each year.

Exotic grasses such as Gamba grass (*Andropogon gayanus*) and mission grass (*Pennisetum polystachion*) were introduced to the tropics to provide feed for livestock. These species can attain fuel loads greater than five times that of native grasses. As a result, fires burning in these fuels can be up to five times more intense than historical fires and are killing native trees previously considered tolerant to fire.

Tussock grasslands

Tussock grasslands are widespread, extending from tropical regions to the Australian Alps. The tussock grasses are perennial species that may grow in nearly pure swards or in association with other annual and perennial grasses to form a continuous fuel bed. Much of our experimental work was carried out in ungrazed kangaroo grass (*Themeda australis*) grasslands. Although these pastures commonly appear to be around 1 m high with inflorescences up to 2 m, the bulk of the fuel in the tussocks occurs below a height of 30 cm. The pastures may be treeless, or associated with woodland or open forest.

Figure 2.3: A spinifex hummock grassland at Rudall River, Great Sandy Desert, Western Australia. Photo: B. Ward.

Over the seasons, a dense mass of dead material can build up within tussocks, often with green shoots dominating the upper area. Most perennial tussock grasses develop coarse unpalatable material with time, and in the tropics they are commonly burnt to remove this.

In temperate climates, tussock grasslands that have not been burnt for long periods can accumulate significant amounts of dead fuel (in excess of 14–15 t/ha) and grassfires will burn intensely even when the outer tussock appears green. They remain flaming longer than annual grasses, and may continue to smoulder for long periods as the dense compacted grass in the centre of the tussock burns out. Under dangerous conditions, fire in these tussocks can be very difficult to extinguish completely with water; often a graded break must be placed around the fire edge to prevent breakaways.

Tussock grasslands of the semi-arid zone, e.g. Mitchell grass (*Astrebla* spp.) country, may be selectively grazed so that by the end of the dry season palatable annuals and ephemerals between the tussocks have been eaten out, leaving bare ground. Under these conditions, fires are limited by lack of fuel and may spread erratically only under the influence of high winds.

Hummock grasslands

The hummock grasslands are characterised by species that grow in a dense clump, forming a substantial hummock or mound after several years. These species range

Figure 2.4: Mallee spinifex association near Pooncarie, NSW. Photo: J. Noble.

from the semi-arid upland three-awned spear grass (*Aristida* spp.) of the Northern Territory through the spinifex grasses – commonly soft spinifex (*Triodia pungens*), hard spinifex (*T. basedowii*) or feathertop spinifex (*Plectrachne schinzii*) – of central Australia and the arid zones, to buttongrass (*Gymnoschoenus sphaerocephalus*) of western and south-western Tasmania.

Spinifex grasses are drought-resisting perennials that form large hummocks 30–60 cm high and 30–100 cm in diameter. The hummocks occupy 30–50% of the ground area, with the interspaces normally bare. After rain or in good seasons, a sparse cover of short grasses and forbs may appear between the hummocks.

Hummock grasses are particularly flammable because the hummocks are composed of a dense core of coarse dead material accumulated over a number of years. A hummock may develop a veneer of fresh shoots on its surface after rain, giving it a green appearance, but it will still burn with surprising ferocity under even the mildest conditions. Some species, such as the buttongrasses of Tasmania, have the capacity to shed water from the outer surface of the hummock and will burn quite readily a day or so after rain. They will burn even when the area beneath the hummock is saturated.

A characteristic of fires in many hummock grasslands is that, before they can spread successfully, the wind speed must exceed a threshold value necessary to produce a flame angle low enough to extend flames from burning hummocks across the intervening bare ground to other hummocks. These fires have little lateral or back spread. Under strong winds they develop a wedge shape from a point ignition, spreading laterally only slowly as the head fire progresses downwind.

Improved pastures

Almost half of Australia is used for sheep or cattle grazing, and in many areas this grazing occurs on what is termed improved pasture. Much of the area was originally woodland but now most trees have been removed and the land sown with a grass species suitable for stock grazing.

Improved pastures often comprise an annual or perennial grass above a lower layer such as subterranean clover. Even under severe conditions, a head fire that burns the standing grass may not burn the compacted clover layer, instead leaving a protective layer of ash on top of it. This can create mop-up problems; when the ash is blown away smouldering material can reignite a fire that can spread across the residual unburnt fuel within the burnt area.

In much of Australia's agricultural land, improved pastures have brought an increase in the fuel load. Very heavy pasture grasses, such as phalaris, have replaced light, fine native grasses, such as kangaroo grass, that originally grew in the woodlands. This change has meant that fires are now much more intense than those of yesteryear and, as a result, can have a devastating impact on the residual trees. Most of these trees are old, and they often have extensive hollows. The fires in south-

Figure 2.5: Improved pasture of phalaris and clover, ACT.

Fuel | 13

Figure 2.6: Wheat stubble near Picola, Victoria. Generally, very little fuel is found between plant rows after harvest.

Figure 2.7: Fire in wheat stubble near Yass, NSW. Sparse surface fuels beneath the stubble causes gaps in the flank fire.

eastern Australia in 1983 burnt down many of the old redgums scattered through farmlands, dramatically reducing the extent of tree cover.

Crop lands

Although less extensive than grazing pastures, crops cover large areas of rural Australia and produce harvests worth millions of dollars in domestic and export markets. Wheat, barley, oats and sorghum are the main crop plants.

Fire behaviour in harvested crop lands can be influenced by the presence or absence of weeds or fine grass beneath the crop. In many cases the ground beneath a crop is bare and post-harvest fires spread primarily through the stalks of the stubble. The bare ground restricts the lateral spread of these fires and, as a result, fires are much narrower than those in more continuous fuels.

In the past, the threat of fire in crop lands following harvest was almost non-existent due to ploughing or deliberate burning of the stubble, resulting in minimal residual fuel on the ground. More recently, the practice of direct-sowing – sowing without prior ploughing or burning – has become popular, resulting in an increase in the amount and continuity of fuel in crop lands. Over a season or two, this has the effect of increasing the intensity of fires as well as their ability to spread consistently.

Comparison of grazed and ungrazed pastures

A progressive change in the nature of the fuel type occurs in pastures subjected to grazing by stock. Often they are well eaten down before they become fully cured, so are shorter than ungrazed pastures of similar species. They are also more compacted. As a result, fires spreading in heavily grazed pastures behave very differently from fires in ungrazed pastures. Similar differences are evident in harvested and unharvested crop lands.

In many parts of southern Australia, fenced road verges are the only areas where grasses can develop to their natural structure. By early to mid-summer, when the grasses are fully cured, it is common to find paddocks grazed to less than 10 cm while the roadside verges have grasses more than a metre high. This tall grass can present special problems for fire suppression if the road surface is the only existing firebreak.

Grassland life cycles

All grasses – annual and perennial – go through an annual life cycle where the plant germinates or produces new shoots, grows, flowers and dies. The general life cycle of an annual grass is shown in Figure 2.8.

The life cycle of all grasses is dominated by rainfall. A wet spring and rain in early summer will produce abundant growth, while an early summer with little or

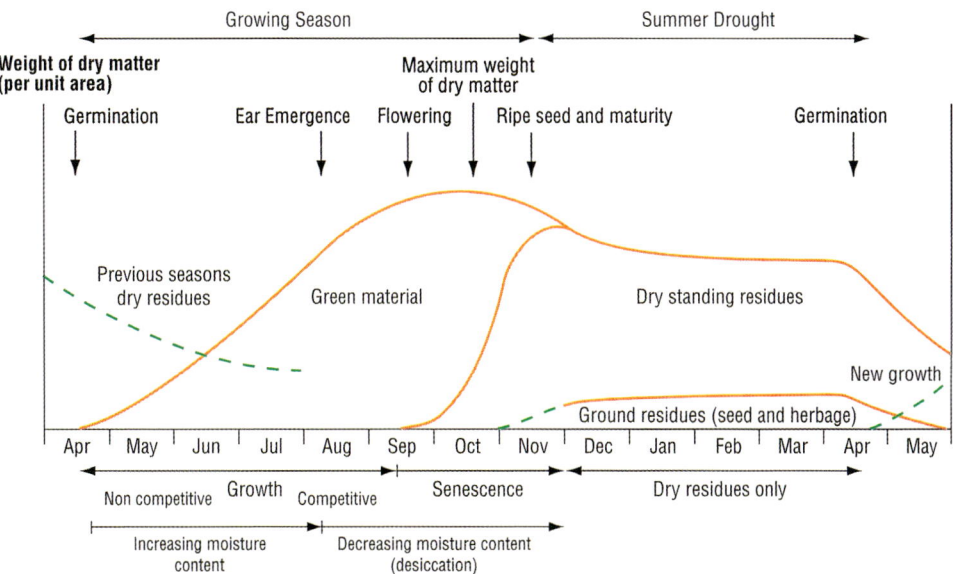

Figure 2.8: Stylised life cycle of annual grasslands near Adelaide. Source: Parrot (1964).

no rain will accelerate the senescence and drying of grasses. A knowledge of grass life cycles and the effect of drought conditions on them is important when estimating the fire danger in a region. Serious grassfires are likely to occur after a good spring season when pastures are fully cured but not yet heavily grazed and trampled

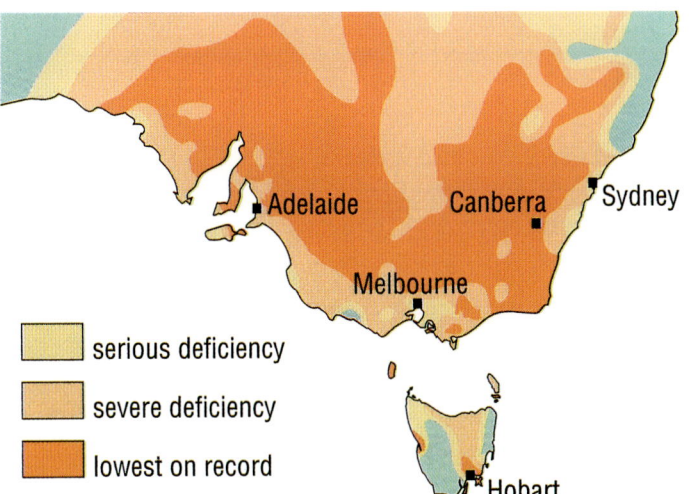

Figure 2.9: Rainfall deficiency map showing the extent and seriousness of the drought over south-eastern Australia in the eight months prior to the Ash Wednesday fires of 16 February 1983. Severe grassfires occurred in areas delineated as serious deficiency. Grasslands where the rainfall was the lowest on record had less significant fires as they carried sparse fuels. Source: Bureau of Meteorology (1984).

by stock. As summer progresses grasslands are eaten down and the chance of a devastating grassfire is reduced.

Different conditions are required to produce severe forest fires. A hot dry summer following a prolonged winter and spring drought is generally required before extensive forest fires can occur. Under these conditions, by comparison, grasslands are frequently eaten-out and carry insufficient fuel to support extensive fires. For example, rainfall for the nine months from 1 May 1982 to 31 January 1983 was the lowest on record through much of Victoria, central South Australia and southern New South Wales (Fig. 2.9). Severe forest fires broke out in East Gippsland, Victoria, in November 1982. Further outbreaks occurred in forested areas of the state with severe rainfall deficiency until mid-March 1983. The total area burnt was in excess of 500 000 ha. Although much of Victoria's pastoral areas had a rainfall deficiency classed as severe or lowest-on-record, severe grassfires occurred only in pockets where the rainfall deficiency was classified as serious but not severe.

3
COMBUSTION OF GRASSY FUELS

The combustion process

Fire is a set of rapid complex chemical reactions that release energy in the form of heat and light. More specifically, the bulk of the energy release occurs as an oxidation reaction, summarised in the combustion triangle shown in Figure 3.1. Each of the three sides – fuel, oxygen and heat – is required for combustion to take place.

In practice the process is a little more complicated, due to both the nature of the fuel itself and the way it can burn. The primary ingredient in all vegetation fuels is cellulose, a natural polymer used to make cotton. When heat is applied the fuel will first dry out, losing any free or bound water from its cellular structure, then proceed to break down. This thermal degradation in cellulose-based fuels will follow two possible paths, depending on the conditions under which the heat is applied. If the heating is slow and the temperature remains below about 250°C, the material will dehydrate and char. This process, which is used to make charcoal, is an exothermic (i.e. releases heat) reaction and is promoted by the presence of water. In its simplest form it can be written as:

$$(C_6H_{10}O_5)_n \longrightarrow 6nC + 5nH_2O + heat$$

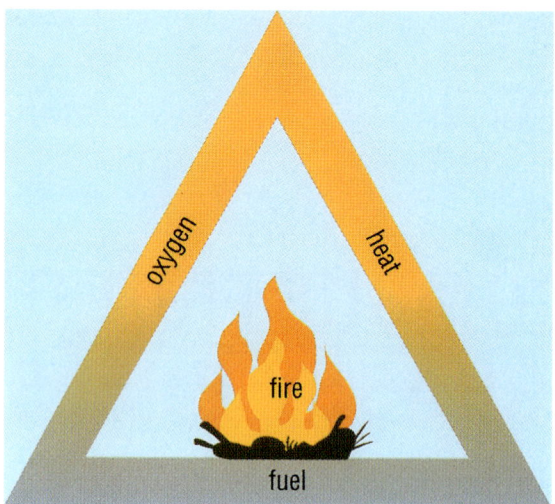

Figure 3.1: The combustion triangle.

However, this reaction (which does not require oxygen) does not produce elemental carbon as written above, but a range of dehydrated carbohydrate compounds that are collectively called charcoal. Depending on the original material and the extent of thermal degradation, some of these charcoal compounds are quite reactive and can subsequently ignite and burn easily in oxygen (e.g. activated charcoal or tinder) while others (charcoal) may require high temperatures and forced air to maintain the oxidation reaction.

If the heating is fast in the absence of water and the temperature is greater than about 250°C, the cellulose will undergo volatilisation. This produces the flammable and highly unstable compound called levoglucosan and is slightly endothermic (i.e. requires heat to continue):

$$(C_6H_{10}O_5)_n + \text{heat} \longrightarrow nC_6H_{10}O_5$$

Levoglucosan can decompose into a wide range of equally flammable and unstable hydrocarbon gases. If the temperature is high enough or a pilot flame is applied, these gases will ignite and combust, combining with oxygen to produce diffusion flames, releasing much heat and a large variety of products that can also subsequently combust with oxygen. The overall reaction for the combustion of cellulose produces carbon dioxide, water and heat. It can be written as:

$$C_6H_{10}O_5 + 6O_2 \longrightarrow 6CO_2 + 5H_2O + \text{heat}$$

This is a highly exothermic reaction. The excess heat produced raises the temperature of adjacent unburnt fuel particles to the level required for thermal degradation, causing a chain reaction. When heating is very rapid, the mixing of volatilised gas with oxygen can be less than ideal, resulting in combustion that is chaotic and incomplete. Large volumes of partially combusted gas (as well as liquid and solid phase products), other hydrocarbons such as tars, and molecular carbon (soot) are produced; these are seen as billows of smoke. Some of the products of partial combustion can be particularly acrid but mostly they are not a threat to health if there is only occasional exposure.

After the flaming has finished, the charcoal compounds formed in the char formation process or residual carbon left over from the incomplete pyrolysis may combine directly with oxygen, resulting in glowing or smouldering combustion that may produce little flame or smoke but significant heat. This reaction can be written in its simplest form as:

$$C + O_2 \longrightarrow CO_2 + heat$$

While grassy fuels consist mostly of cellulose, the fuel bed contains other compounds, elements and minerals. When combustion is complete, these remain as the residual ash.

Volatilisation and char formation will occur concurrently to varying degrees at all locations around the fire perimeter. The completeness of the combustion of the products of thermal degradation dictates the appearance of the resulting smoke. Ideal mixing of air with the flammable products will result in little smoke, in either flaming or glowing combustion. If combustion is restricted in any way, denser smoke will be produced.

Diffusion flames

Diffusion flames result from the combustion of flammable gases that have not been mixed with oxygen prior to ignition. Volatilising solid fuel will form a bubble of unburnt flammable gas that will mix with oxygen on the surface of the bubble through the process of molecular diffusion. Once this mixture ignites, it will burn along the interface between the fuel-rich/oxygen-free region and the fuel-free/oxygen-rich zone, known as the reaction zone (Fig. 3.2), much as the flame of a cigarette lighter or a candle burns. The reaction zone will also comprise burnt products once fuel and oxygen molecules have reacted. The build-up of burnt product can inhibit the mixing of the fuel gas and oxygen, resulting in less-than-complete combustion as the temperature of the fuel gas drops below that required for combustion.

The action of the buoyant nature of the heated fuel gas and burnt products causes the interface between the fuel gas and the air to ripple, deform and become convoluted, increasing the area of contact and thus the area of the burning reaction zone (Fig. 3.3). These flames are known as turbulent diffusion flames and can give the appearance of a solid sheet of flame; in reality they comprise a number of folds in the reaction zone.

The temperature within the reaction zone is around 1600°C. However, because the reaction zone is sandwiched between cooler unburnt fuel gas and much colder

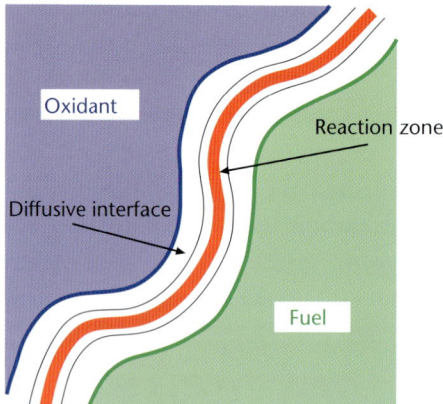

Figure 3.2: The nature of a diffusion flame, as found in a grassfire, is that the burning reaction zone is a thin zone separating unburnt fuel and oxidant (air). Temperatures in this zone can reach 1600°C.

ambient temperature air, the mean temperature of an entire flame is something much less. Thermal measurement of flame fronts in grassfires has found that the temperature of a flame ranges from around 300°C at the tip to around 1100°C at the base if the flames are more than 6 m high.

Figure 3.3: A single reaction zone is almost transparent. Turbulence causes layers of reaction zone to overlap. Multiple layers radiate heat more effectively, as indicated by the regions of brightness in the photograph.

Table 3.1: Heat yields of grassy fuels determined with a flow calorimeter

Grass species	Moisture content (% oven-dry weight)	Heat yield (kJ/kg)
Phalaris tuberosa	6	16 000–16 300
Phalaris tuberosa	10	13 700–13 900
Phalaris tuberosa (partially cured)	47	11 500
Themeda australis	10	14 500–14 900
Themeda australis	8	16 900–17 800
Eriachne spp.	11	15 100–16 700
Eriachne spp.	10	15 200–18 500
Sorghum intrans	10	16 900–17 600
Sorghum intrans (stems only)	13	17 400–18 600
Sorghum intrans (leaves only)	13	16 000–16 500

Heat yield from grassy fuel

The heat released from the complete combustion of fuel is called the heat of combustion, and is expressed in terms of kilojoules per kilogram (kJ/kg) of fuel. Combustion is rarely complete under field conditions, however, and some energy remains in partially burnt combustion products and residues. Hence the heat released (the heat yield) is less than the heat of combustion. The heat yield is quite variable because it depends on the way the fire burns, as discussed in 'Fire types' below. Typical heat yields for grasses are given in Table 3.1.

In general, heat yields are lower at higher moisture contents because heat is required to remove the water and raise the fuel to the flame temperature, and because combustion of moist fuels is less complete. Heat yields are used to calculate fire intensity.

Moving grassfires

The process of combustion is the same for all vegetation fuels, whether forest or grassy fuels. The rate of combustion is evident in the height of the flames – the faster combustion proceeds, the higher the flames – and depends primarily on the size of individual particles and their compaction in the fuel bed. All pasture grasses have a similar range of particle sizes, but height and bulk density are likely to differ within a sward. Even so the range is not great; natural swards carrying fuel loads of 2–6 t/ha have a bulk density of about 1 kg/m^3, and grass compacted by mowing or grazing may have an average bulk density of 3 kg/m^3. The greatest distinction is between forest and grassland fuels. The leaf, stick and bark material of forest fuel is

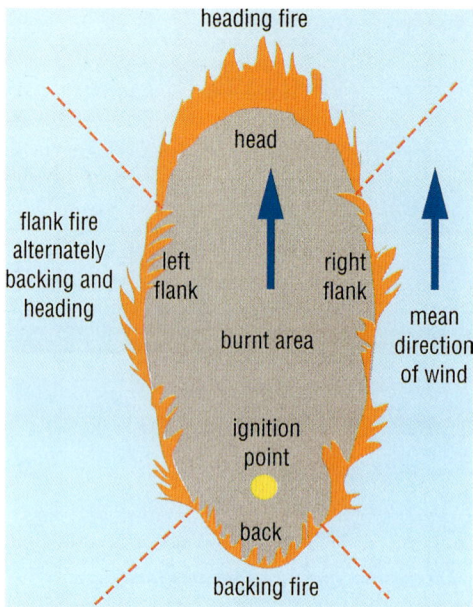

Figure 3.4: The parts of a moving fire. Flames of the heading fire lean into unburnt fuel; flames of the backing fire lean over burnt ground; flames of flank fires may lean over either the unburnt fuel or the burnt ground depending on local fluctuations in the wind direction.

larger than grass particles and the litter layer has a bulk density of around 50 kg/m³. Thus combustion rates are slower in forest fuels than in grasslands, and flames persist longer at any one spot.

However, important differences are evident in patterns of combustion of grassy fuel beds. These patterns depend on how the grass is ignited and the speed with which fire spreads through the fuel bed, which primarily depends on whether the fire is burning with or against the wind.

Fire types

There are three types of fire, each related to the orientation of the fire's edge with respect to the wind. In moving fires, different fire types will occur at different positions along the fire perimeter. Consider a fire that starts from a point under a steady wind. It will form a roughly elliptical shape, with all sections of the perimeter fitting into one of the three classes – heading, backing and flanking fire (see Fig. 3.4).

Heading fire

A heading fire is one where the flames are blown towards the fuel (Fig. 3.5). The fuel bed is ignited at the top and the fire progressively burns down into the lower layers. This pattern of burning can be quite inefficient, particularly under severe weather conditions.

Figure 3.5: Experimental fire A091, NT, 4 August 1986. Black smoke indicates incomplete combustion. Flashes of flame indicate detached envelopes of combustible gas.

Initial spread of heading fires burning dry aerated grass may produce little smoke, which indicates that combustion is complete. A little black smoke may be produced because the fuels are pre-heated and ignite so rapidly that the large volume of flammable gases does not mix sufficiently with oxygen to permit complete combustion, thus partially combusted products and free carbon (soot) escape from the flame zone.

As the fire progresses the residue of incompletely combusted material from the upper part of the fuel bed can be deposited on the lower unburnt fuels, restricting the mixing of oxygen with the burning fuel and decreasing combustion efficiency. This can produce a large volume of smoke consisting of hydrocarbon compounds, commonly called tars, which mixes with the soot from the burning of the upper layer and can appear as dense black smoke. The lower stalks and compact layers at the bottom of the fuel bed, such as compacted clover, that are coated with the black carbonised residue may remain unburnt – seemingly insulated by the deposited material.

Backing fire

Backing fires are those that move into the wind, with flames leaning over the burnt ground (Fig. 3.6). These fires ignite the fuel bed at or near the base and burn slowly but very efficiently, leaving little residue in the form of partially burned material or carbon. They mostly leave a fine white ash. Because of their efficient combustion, backing fires produce less smoke than heading fires. In areas where there are

Figure 3.6: Experimental fire A011, NT, 30 July 1986. Flames burning from the base of the fuel bed are leaning over burnt ground.

concerns about air pollution, backing fires can be used to burn crop residues with minimum smoke emission.

Flanking fire

In a flanking fire, the fire edge is generally parallel to the direction of the wind (Fig. 3.4). This means the flames will lean, more or less, along the flank. Gusts of wind picking up a flanking fire will cause high flames to travel in waves down the flank. Viewed side-on, it will appear as though the flames are spreading at very high speed, perhaps equivalent to the gust speed.

The major characteristic of a flanking fire in grass fuels is that it will become a heading fire and a backing fire in response to changes in wind direction, and will have attributes of each (see Fig. 3.7).

At different times in a fire's development, heading fires, flanking fires and backing fires may occur at any location around the fire perimeter, depending on fluctuations in wind direction. The distinct differences in the way these three types of fires burn are useful characteristics to consider in post-fire analysis of fire spread.

Flame zone characteristics

Flame height, flame length, flame angle

The height of the flames – their vertical height above the ground (Fig. 3.8) – is the most obvious characteristic of a fire. While this is readily observed, and often

Combustion of grassy fuels | 25

Figure 3.7: Flank fire in grassy woodland, NT, showing both heading and backing fires on sections of the flank.

reported or recorded by visual observation, it is difficult to obtain consistent readings or a high degree of precision. This is because the diffusion flames of a grassfire are the result of flammable gases emitted from the fuel bed combining in a turbulent way with oxygen in the surrounding air. Billows of gas may not

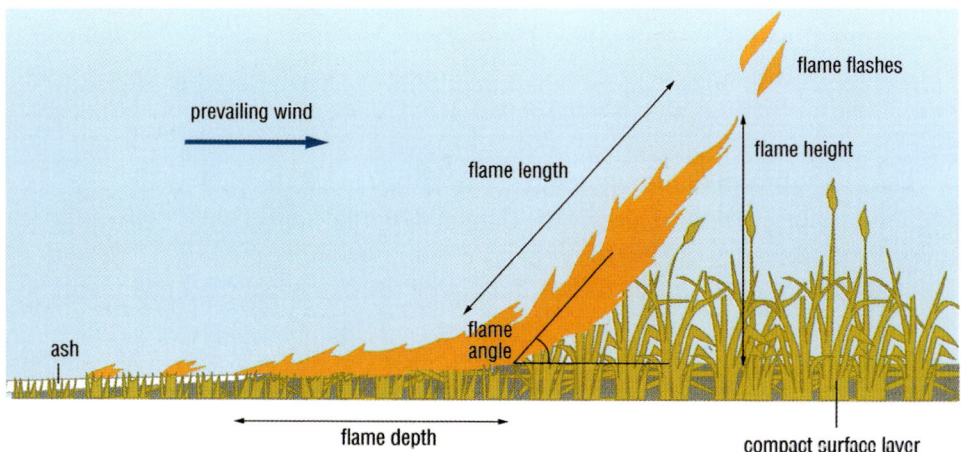

Figure 3.8: Stylised cross-section through a heading fire. The tall flames at the front result from the combustion of the standing grass. The depth of flame is largely determined by the amount of material in the compacted surface layer.

Figure 3.9: Experimental fire B052, NT, 19 August 1986. Flame height 4 m; flame length 6 m; flame angle 45°; height of highest flash 7 m; length of flash 10 m; rate of spread 5.5 km/h.

burn completely as they rise from the fuel bed, or bubbles of gas burning on the exterior may become detached from the flames below and burn out as they are carried aloft.

The convective forces of the buoyant combustion gases and the wind establish a dynamic balance that results in flames of some average length being inclined over the fuel bed. However, fluctuations in wind velocity, turbulent mixing and variations in fuel structure and composition cause wide fluctuations around the mean length. Despite this, flames do achieve a characteristic average height that depends on the nature of the fuel bed and the prevailing weather conditions.

The most consistent estimates of flame height, taken vertically from ground level, come from estimates of the average height of the tops of the flames as viewed from a distance and integrated for some time and some distance around the perimeter. Such estimates ignore occasional high flame flashes and gaps in flames created by turbulence (Fig. 3.9). Nevertheless, we should not expect a great degree of precision; flame height estimates in steps of 0.25, 0.5. 1, 2, 4 and >4 m should be adequate for most purposes.

Flame length and flame angle are more difficult to estimate. They also fluctuate and are variable over the depth of the flames. The angle of the tallest flame, which reflects the buoyancy of the most vigorous combustion, is probably the easiest to estimate, but the angle at which the front face of the flames subtends to the fuel bed is probably more important for the physics of heat transfer. For practical purposes, estimates of flame height and the most obvious angle of the flames, without worrying too much about its location, are useful fire descriptors.

Residence time

Residence time is the period during which flames remain burning over one spot on the ground. Depending on its thickness, an individual particle of grass remains flaming for 2 seconds or less. The fuel bed takes longer to burn, the length of time depending on the fuel load and its compaction. Heavy pastures commonly have a residence time of 10–15 seconds, while fine light pastures have a residence time of 5 seconds or less.

The short residence time of grassfires means they are extremely responsive to changes in wind speed and direction. An alteration in wind direction can change a flank fire to a head fire, which within 15 seconds can spread at its maximum speed. Conversely, a lull in the wind can, within 5–10 seconds, reduce a fast spreading head fire to a backing fire with low flames. In tropical areas, strong thermal activity can cause large variation in wind speed and therefore large variation in fire behaviour. This behaviour has been described as erratic, but it is merely a reflection of wind variability.

Residence time is one of the most important characteristics of the flame zone. It determines the depth of the flame zone (Fig. 3.8) and the length of time any object overrun by a fire will be immersed in flame. It is the main cause of the differences in behaviour between grassfires and fires in other fuel types such as forest or scrublands.

Flame depth

Flame depth is the distance behind the fire front covered by continuous flames (Fig. 3.8). The measurement does not include isolated flaming tussocks or logs which may persist for a considerable time. Flame depth is a useful fire behaviour characteristic, but can be difficult to observe from ground level when flames are tall. Flame depth (d_f) in metres can be calculated as the product of the fire's rate of spread (R) in metres per second and the residence time (t_r) in seconds:

$$d_f = R \times t_r$$

For example, a high-intensity grassfire travelling at 18 km/h (5 m/s) in a fuel bed with a residence time of 10 seconds will have a flame depth of 50 m.

Firefighters dressed in normal clothing cannot expect to pass through flame depths of more than a few metres and survive. However, because grassfire residence times are short a sudden drop in the rate of spread caused by a lull in wind speed or a change in wind direction means flame depths will also reduce within a few seconds. It then becomes possible to step across low flames onto burnt-out ground. If the smoulder time of the fuel bed is long, it will be necessary to move a considerable distance into the burnt area before conditions are safe and comfortable. In woodlands and open forests where there may be downed woody material, that distance is even greater as these larger fuel components have significantly longer residence times and are hazardous sources of radiant heat sufficient to cause severe burns long after the fire front has passed.

Smoulder time

The length of time that fuel will smoulder after the flames of a head fire have passed (the smoulder time) depends on the structure and compaction of the lower layers of

Figure 3.10: Experimental fire C064, NT, 18 August 1986, in compacted fuels. Rate of spread 4.8 km/h, flame depth 8 m, residence time 13 seconds, smoulder time approximately 1 min. The block is 100 × 100 m.

the fuel bed. Often, much of the smoke produced by a grassfire comes from smouldering combustion behind the flame front.

In fast-moving fires a zone of smouldering combustion can be observed in an arc behind the flaming zone (Fig. 3.10). The smoulder time in perennial *Themeda* pastures in northern Australia is around 60 seconds. Annual pastures and crop lands with little compacted material on the surface have very short smoulder times. On the other hand, the cores of very large tussocks may continue to smoulder for several hours or even days. If the oxygen supply is limited, e.g. in the core of a large tussock or around underground plant parts, smouldering combustion may be reduced but continue efficiently and produce very little smoke. Such fires can be difficult to detect visually and difficult to extinguish.

Heat transfer

Heat from the combustion process in the fuel bed, as manifested by flames, is transferred by radiation, convection or conduction. Generally, we consider that conduction into the ground is very small and plays little part in the propagation of grassfires. Measurements of radiant and convective heat transfer have shown that radiation accounts for about 20% of the total transfer of the potential energy of the fuel when burnt; the remainder is transferred by convection. However, convective heat is mostly buoyant – it is carried upwards away from the ground – so what firefighters mostly feel is radiant heat.

The potential radiation flow, or flux, from a continuous grassfire flame – the radiation flux we would be subjected to if totally surrounded by tall deep flames – is around 100 kW/m^2. As a comparison, peak sunlight on a clear, bright summer's day is around 1 kW/m^2. The threshold of pain for radiant heat on bare skin for most people is around 2 kW/m^2. The radiation actually felt by a person or object confronted by an approaching fire is much more difficult to determine. The radiation flux from a line of fire is roughly inversely proportional to the distance from the fire – radiation from a point source is inversely proportional to the distance squared. In practice, however, radiation flux depends on the shape of the flame (angle and length), the curvature of the fire front, the distance of the object from the fire and what may intercept the radiation across that distance. Calculation of the radiant heat received at any point in front of a fire is generally impractical given the highly variable nature of moving flames, although there are methods that incorporate a number of approximations of the fire's geometry.

Fireline intensity

Fireline intensity is a calculated number that represents the rate at which heat is released from a lineal segment of the fire perimeter. Expressed in kilowatts per metre of fire edge (kW/m), it is given by the equation:

$$I = H \times w \times R$$

where H is the heat yield of the fuel burnt (kJ/kg), w is the amount of fuel consumed (kg/m^2) and R is the rate of spread (m/s). If we assume that H = 18 000 kJ/kg (see Table 3.1), a quick estimate of fireline intensity can be obtained from:

$$I = 500 \times w \times R$$

where w is the fuel consumed in t/ha and R is in km/h.

The rate of spread of each metre of the fire perimeter varies depending on its location on the perimeter. At the head, where the rate of spread is greatest, the fireline intensity is greatest and at the back of the fire where the rate of spread is least the intensity is least. Fires are usually characterised by the intensity of the head fire. A theoretical illustration of the relative intensity calculated for each metre around the perimeter of a grassfire is shown in Figure 3.11. In reality the intensity of the flank fire can range from the intensity of the head to the intensity of the back fire as it responds to fluctuations in wind direction as discussed in 'Flanking fires' above. Additionally, the heat yield realised at the flanks and rear of the fire will be different from that at the head.

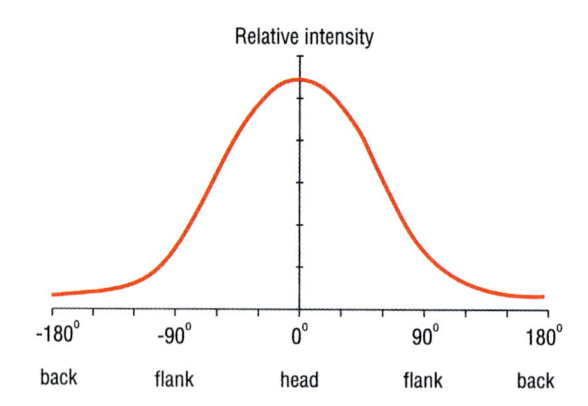

Figure 3.11: An illustration of the relative fire intensity around the perimeter of a grassfire burning under a steady wind. Source: After Catchpole *et al.* (1982).

The intensity of a grassfire may range from a 10 kW/m backing fire in light fuels to 60 000 kW/m at the head of a very fast wildfire. Fire intensity is useful when comparing fires in the same fuel type, but should not be used to compare fires in different fuel types, e.g. those in grass with those in forest fuels. Because of the very different combustion characteristics of the two fuels, the behaviour and flames of a forest fire will be very different from those of a grassfire of the same calculated intensity.

4
FIRE BEHAVIOUR

Fire behaviour is everything a fire does. It covers the way a fire ignites, builds up or grows, the rate of spread, the flame front characteristics and all other phenomena associated with the moving fire front. In this chapter we discuss in general terms the factors that affect the behaviour of a grassfire from ignition through its development until it attains its potential rate of spread. Chapter 5 covers the main factors used to predict a grassfire's rate of spread across the landscape, while Chapter 6 examines the variability of fire spread and, in particular, its dependence on wind speed variability.

Ignition

The ease of ignition of grassy fuels depends primarily on the moisture content of the dead grass. At moisture contents above 15%, only a sustained flame can cause ignition. Ignition becomes progressively easier as the moisture content of dead fuel decreases; below a moisture content of 6% very small embers or hot particles are capable of igniting grassy fuels. Wildfires can then be started by factors that do not normally cause ignition, such as glowing carbon particles from defective exhausts, hot metal sparks from clashing power line conductors, grinding operations and metal striking rock during the operation of slashers or bulldozers.

Recently dead grass is not readily set alight by small embers, particularly if it is standing upright and there is little material on the soil surface. Partially decomposed grassy material is more likely to ignite because the embers can make good contact with the fuel.

Wind at the fuel level may aid or hinder ignition. Wind will assist ignition from glowing combustion such as from cigarette butts or woody embers, but will hinder ignition by metal sparks because they hasten cooling of the fragments. Once glowing combustion has begun, high wind will increase the combustion rate and promote flaming combustion. Of course, following any successful ignition, high winds result in rapid development and fires become difficult to put out even when they are small.

Other reported ignition agents include spontaneous combustion, glass bottles, glass fragments and high-pressure aerosol cans. Spontaneous combustion in natural fuels is most unlikely, although it does occur in special situations such as silage pits, wet-baled hay and in farm buildings where rags soaked in linseed oil are known to be an important catalyst.

Laboratory trials have shown that clear glass bottles containing enough water to form a lens can ignite fires under controlled conditions by focusing the sun's rays onto a fuel bed. Under field conditions, though, it is highly unlikely that a bottle or glass fragment will form a lens of the correct focal length and orientation to concentrate sunlight sufficiently and long enough to start a fire. Similarly, while it is theoretically possible to reflect and concentrate the sun from the concave base of an aerosol can, in practice this is highly unlikely. Considering the large amount of glass and can litter along the roadsides, the chances of ignition by these agents must be very small indeed.

Fire growth

All fires increase their rate of spread after ignition until they reach a quasi-steady rate for the prevailing weather conditions. We use the term 'quasi-steady' because the instantaneous rate of spread is always responding to fluctuations in wind speed. The quasi-steady, or average, rate of spread will remain the same while the average wind speed remains constant. At a constant fuel moisture content, this rate depends on the wind speed and the width of the head fire.

Figure 4.1a shows the growth of a fire lit at a point and mapped at 2 minute intervals. Figure 4.1b shows a pattern of forward spread that reflects both short-term (2 minute) fluctuations in wind speed and the build-up of the fire through several quasi-steady steps towards its potential rate of spread for the prevailing conditions. During the first 20 minutes, the fire remained narrow with a head fire 20 m wide and a quasi-steady rate of spread of 0.4 km/h. Then a shift in wind direction increased the width of the head fire to 40 m and the quasi-steady rate of spread rose to 0.8 km/h. Three more wind shifts occurred during the fire, and each time both the width of the head fire and the quasi-steady rate of spread increased significantly. This is despite the fact that the average wind speed remained more or less constant, increasing only slightly during the life of the fire. If a fire is not affected by shifts in

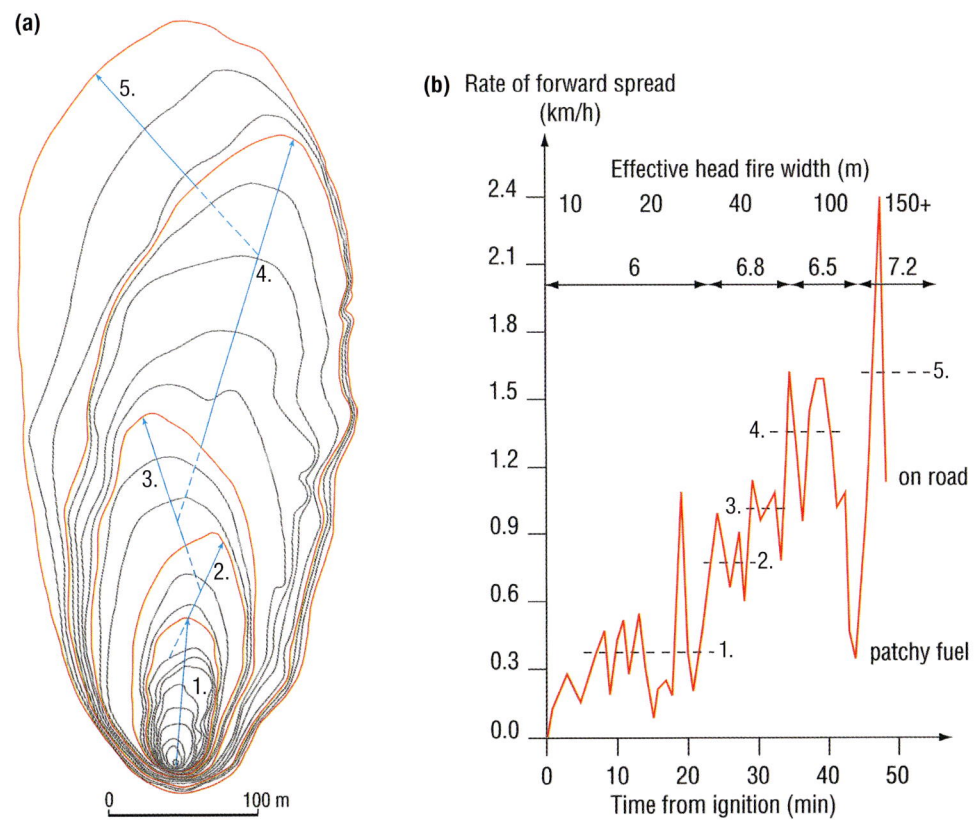

Figure 4.1: (a) Growth of experimental fire No. 50 in open forest at Gunn Point, NT, starting from a point and responding to changes in wind direction. The isochrones represent the perimeter at 2 minute intervals. The arrows represent the direction of wind shifts during the fire. (b) The fire shown in (a) increased in speed and width in a stepwise fashion. Each progressive increase in the width of the head fire was caused by a shift in wind direction. The dashed line (– – – –) represents the approximate quasi-steady rate of spread for each width class. The mean wind speed in km/h at 2 m above the ground is shown as ←6→.

wind direction, it will maintain a narrow head fire and thus may not reach its potential maximum rate of spread.

The head fire width required for a fire to attain its potential maximum rate of spread increases with increasing wind speed. Figure 4.2 shows the quasi-steady rate of spread that will be reached for a particular width of head fire at wind speeds of 7, 14 and 21 km/h. At a low wind speed of 7 km/h the fire has reached its potential rate of spread (1.7 km/h) when the head fire width is around 30 m. At a wind speed of 14 km/h, the head fire width must exceed 100 m before the fire reaches its potential rate of spread, while at a high wind speed of 21 km/h the head fire width must be greater than 150 m. Under very strong winds fires will continue to increase their rate of spread until the head fire reaches a width of around 200 m.

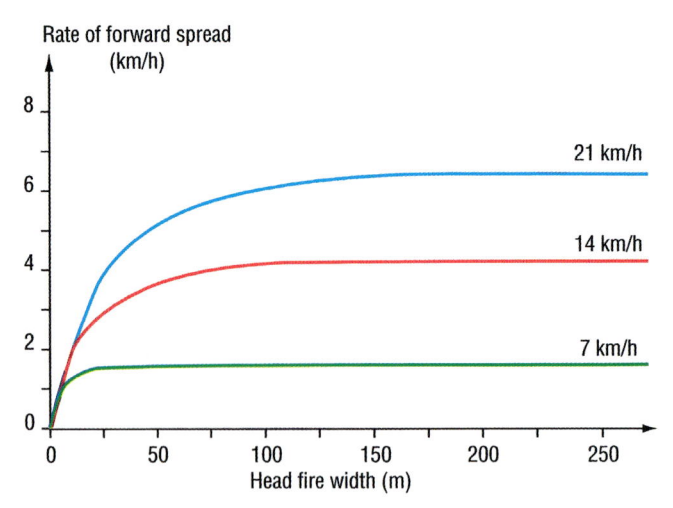

Figure 4.2: Relationship between rate of spread and head fire width for fires burning in open grassland with a fuel moisture content of 4.5% and wind speeds of 7, 14 and 21 km/h.

The time a fire takes to reach its potential rate of spread for the prevailing conditions can vary greatly. Fires developing under unstable conditions, such as a hot northwesterly wind with frequent and substantial shifts in direction, will increase their head fire widths quickly, and the time needed to attain the potential maximum rate of forward spread will be short. On the other hand, fires developing under a stable wind with few shifts in direction, such as a sea breeze or after a cool change, will have a narrower head fire and take considerably longer to reach their potential maximum rate of forward spread. During experiments, the time taken for a fire starting from a point to reach its potential rate of spread has varied from 12 minutes to more than an hour. Generally, most fires reach this rate of spread within 30 minutes, but because the time taken depends on fluctuations in wind direction there are no accurate relationships that can be used to describe the growth of a fire with time.

If a fire is spreading slowly with a narrow head, a 90° wind shift will turn the flank fire into a head fire. This fire will reach its potential rate of spread almost immediately and may travel at up to three times the previous speed. Firefighters should be most wary of the changes in fire behaviour associated with wind shifts when suppressing flank fires.

Likewise, fires that are lit from a long continuous line will reach their maximum rate of spread very quickly. This should be borne in mind any time fires are lit to run with the wind, for example when stubble burning or when burning-out to secure the perimeter of a wildfire. Line fires are often used to widen a firebreak in advance of a wildfire. These line fires may be just as intense as the approaching wildfire and will be just as difficult to hold. It is better to use a series of point ignitions that do not develop to their potential rate of spread before they reach the break (see Figs 4.3 and 8.16) and burn the bulk of the area by a flank fire.

Figure 4.3: Experimental fire F24, NT, 20 August 1986. Three fires lit at the same time at a point, along a 50 m line and along a 100 m line showing how the rate of spread is influenced by the width of the fire front. If the head remains narrow the fire will spread more slowly than a larger fire with a broad head.

Fire shape

Wind speed largely determines the shape of a free-burning fire, especially in the early stages of its growth. The stronger the wind, the narrower and more elongated the fire. Conversely, the lesser the wind the wider the fire. Figure 4.4 shows the relationship between wind speed and a fire's length-to-breadth (L:B) ratio. A fire burning in no wind will have a circular perimeter (L:B = 1.0).

This relationship remains true as long as the mean wind direction remains constant; if the wind direction varies the L:B ratio will be considerably less. In discontinuous fuels such as hummock grasslands and eaten-out pastures, the L:B ratio will be much greater due to restricted flanking and backing spread.

Fuel characteristics

Fuel continuity

Fuel continuity is the major fuel characteristic influencing fire spread. Fuel continuity is the term used to describe the extent to which the surface of the ground is covered by fuel. A fuel that has 100% ground cover is continuous, while one that has

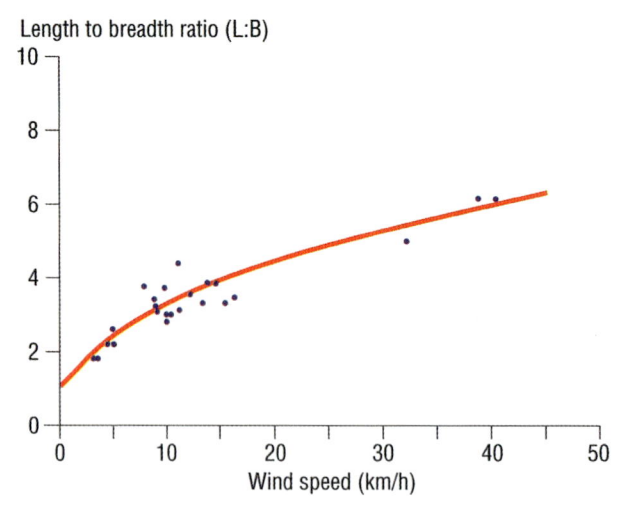

Figure 4.4: The shape of a free-burning grassfire related to wind speed (U) in the open at 10 m. The length-to-breadth ratio (L:B) = $1.1U^{0.464}$. Source: Luke and McArthur (1978).

bare ground between clumps is termed discontinuous. Grasslands that are discontinuous (see Figs 2.3 and 2.4), either because of the natural distribution of the grasses (e.g. hummock grasslands) or because of very heavy grazing (particularly in some tussock grasslands), will not carry a fire until the wind speed exceeds a particular threshold value.

Fuel height

Fuel height dramatically influences the height of flames and hence the difficulty of suppression, but has only a relatively small influence on a fire's rate of spread. For example, experimental fires travelling at the same speed had flame heights of 3 m in tall ungrazed pasture and 0.5 m in mown pasture (Fig. 4.5). The relationship between flame height and rate of spread in these pastures is shown in Figure 4.6.

Theoretically, we would expect fuel height to influence a fire's rate of spread. However, we could not find a significant influence of fuel height on rate of spread in ungrazed natural grasslands. We did observe, however, a reduced rate of spread when the grass was mown or grazed. The average difference between rate of spread in tall natural pastures and in mown or grazed pastures was only 20% over a wide range of wind speeds.

Part of the problem may lie in the difficulty of defining fuel height. The density of a grass sward is greatest close to the ground and decreases with height. Thinly spaced leaves and flower heads at the very top of the sward delineate the height of the pasture, but these fuels are too widely spaced to influence fire spread. While there must be some height at which the average density of the sward can be determined and related to fire spread, definition and measurement of this height is practically impossible.

Figure 4.5: Fuel height and fuel load affect flame height but not rate of spread. Both fires are travelling at 3 km/h under a mean wind speed at 10 m of 22.5 km/h. Top: Experimental fire A071 – flame height 3 m, fuel height 24 cm, fuel load 4.9 t/ha. Bottom: Experimental fire A073 – flame height 0.5 m, fuel height 10 cm, fuel load 2.4 t/ha.

Because the flames from tall grasses are higher and often show violent movement, fires in tall grasses appear to be more erratic and to travel faster than those in shorter grasses. However, we have not been able to measure a significant difference in rates of spread in natural pastures with very different heights and densities.

Fuel load

Fuel load by itself does not influence rate of spread of grassfires. Previous publications have reported that fuel load directly influences a fire's rate of spread, but we have found that this occurs in grasslands only if changes in fuel load also reflect changes in fuel condition, particularly changes in fuel continuity. Fuel load by itself

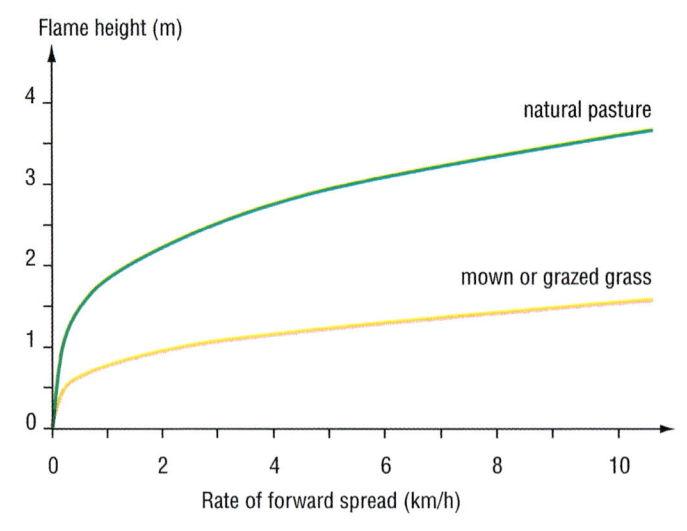

Figure 4.6: Relationship between flame height and rate of spread in natural or ungrazed pastures (0.5–0.8 m tall) and mown or grazed grasslands (<0.25 m tall).

will influence flame height, flame depth and fire intensity, and thereby the difficulty of suppression.

The reason for this apparent contradiction is that in the past fuel load was used to quantify the condition of the pasture, i.e. the extent to which it had been grazed or eaten-out. Heavy fuel loads generally imply swards of tall standing grass and light fuel loads are normally associated with grazed or eaten-out pastures. In these situations it is the condition of the pasture, i.e. the degree of grazing, that determines a fire's rate of spread. If the pasture conditions are the same and are continuous it really doesn't matter if the fuel load in each pasture is heavy or light.

Fires spreading across a landscape that is crossed by roads or other barriers are likely to spread more slowly in short standing grass than in tall standing grass. This is because barriers are more effective against short flames and fires in short grass will be delayed for longer, producing an overall slower rate of spread. The speed of the fire between the barriers will be the same in both pasture conditions.

Other factors, such as the fineness of the grass stalks, may influence residence time. Laboratory studies suggest that fuel fineness also strongly affects rate of spread, but field studies in Australian pastures have shown little obvious influence.

Fuel moisture content

Live fuel moisture content or grass curing state

The curing state of a grassland, expressed as the fraction of dead material in the sward, has a major effect on fire spread.

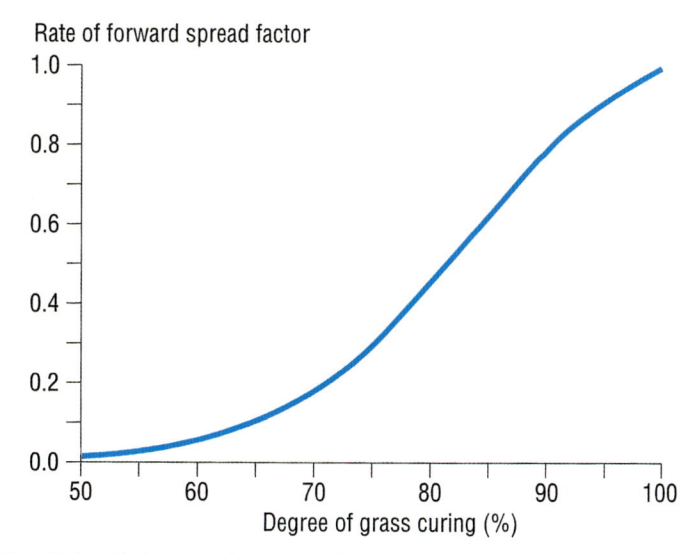

Figure 4.7: The relationship between the degree of grass curing and the rate of forward spread factor. This factor is a fractional multiplier of the rate of spread in fully cured grass. For example, at 50% curing, fires will not spread; at 70% curing, fires spread at 0.2 times their rate of spread in fully cured grass.

The relationship between the rate of spread in uniformly cured pastures and curing is shown in Figure 4.7. Fires will not spread when grasslands are less than 50% cured. The greatest rate of change in the effect of degree of curing on fire spread occurs when grass is between 75% and 90% cured. When grasslands are more than 95% cured they have almost reached their full potential for fire spread.

Dead fuel moisture content

The moisture content of dead grasses under field conditions can range from around 2% to 35% of their oven-dry weight (ODW). Above 35% ODW, the grass fibres are saturated and there is free water on or within the stalks. In the field, fires will not spread under light winds when the dead fuel moisture exceeds about 20% ODW. The precise value of a fuel's moisture content of extinction – the moisture content at which the fuel will not burn – can be difficult to determine experimentally because humidity, and hence dead fuel moisture, often increases rapidly when night falls and fires go out.

Fires under the influence of strong winds may continue to spread at dead fuel moisture contents that would normally prevent spread when winds are light. Windy conditions (>10 km/h at 10 m) create better mixing of air at the grass surface, and dew does not form readily. As a result, fires may continue to burn up to a dead fuel moisture content of 24% (Fig. 4.8).

Hummock grasslands have much higher moisture contents of extinction than other grasslands. Spinifex grasslands are difficult to ignite at fuel moisture contents

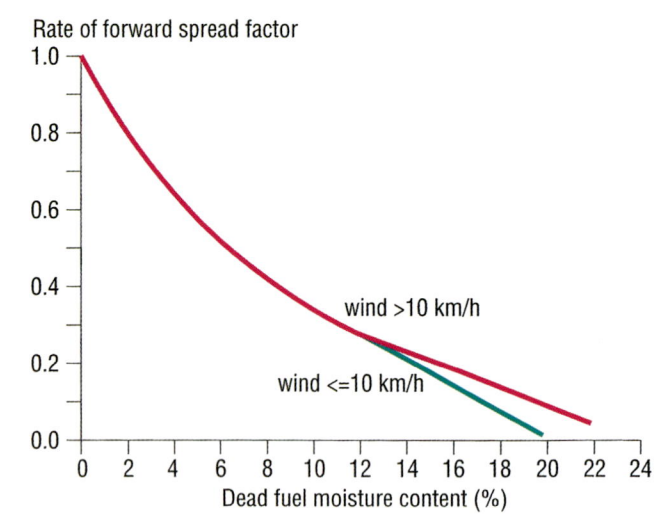

Figure 4.8: The relationship between dead fuel moisture content (DFMC) and rate of forward spread factor. This factor is a fractional multiplier of the rate of spread in fuels with a DFMC of zero. In light winds fires cease to spread at a DFMC of 20%, while under windy conditions they continue to spread up to a DFMC of 24%.

in excess of 35%, although in this case the fuel moisture content measurement included both live and dead material.

Wind speed

Wind is the most dynamic variable influencing grassfire behaviour. Wind speed fluctuates widely over short periods, and varies with height above the ground. When relating wind speed to fire spread it is necessary to specify the height at which the wind is measured and the period of measurement; the latter must be the same for both wind speed and fire spread.

Backing fires

Wind speed has a dramatic effect on heading fires but little effect on fires backing into the wind. The backing rate of spread is largely unaffected by increasing wind speed, but is influenced by fuel moisture (Fig. 4.9). The backing rate of spread is also strongly affected by minor changes in fuel continuity that have no influence on the forward rate of spread.

Backing fires in light pastures can be blown out by very high winds. During the 1983 Ash Wednesday fires in the south-east of South Australia, winds associated with the frontal change, which averaged 70 km/h with gusts to 100 km/h, extinguished many kilometres of the western flank of fires in light pastures. Backing fires in tall ungrazed grasses were not extinguished and continued to burn back into the wind, probably because combustion at the base of the fuel bed was sheltered from the wind. Various blowing devices have been suggested for fire suppression – a

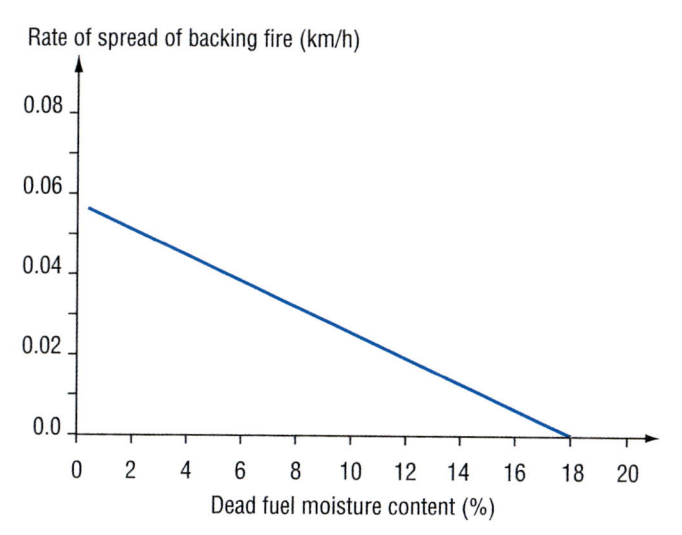

Figure 4.9: Relationship between backing rate of spread and dead fuel moisture content for fires in continuous fuels of Sorghum intrans and other tropical grasses. The rate of backing spread was unaffected by wind speed within the experimental range of 3–20 km/h.

handheld leaf-blowing device is used in China – but these have limited utility in sparse pastures and are largely ineffective in dense tussock grasses because the wind only accelerates glowing combustion deep in the tussock.

Heading fires

The relationship between wind speed and the rate of forward spread of heading fires is shown in Figure 4.10. Field observations of wind speed and rate of spread over

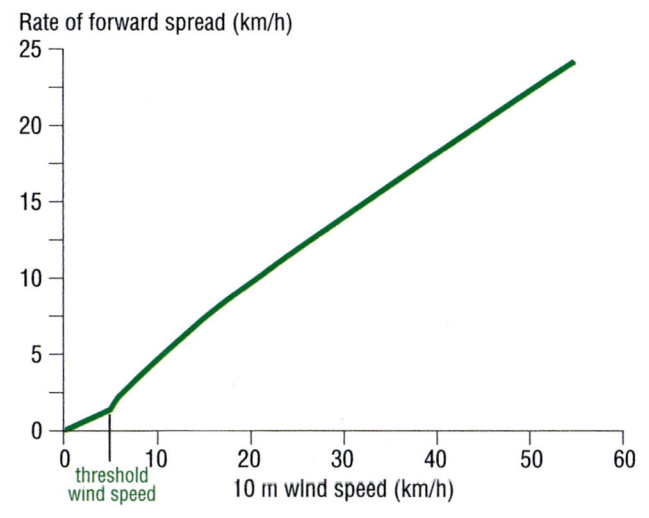

Figure 4.10: Relationship between average wind speed at a height of 10 m and rate of forward spread. The threshold wind speed is the speed at which the fire spreads as a continuously heading fire. Above this speed the relationship is slightly curvilinear; increase in rate of spread lessens slowly with increasing wind speed.

short periods may appear to vary considerably from the values derived from this relationship. This is because the speed of the wind passing the anemometer is often different from that blowing at the front of the fire (see Chapter 6).

Much closer agreement will be obtained when average values of wind speed and fire spread are taken over a longer period. To predict fire behaviour, therefore, wind speed measured at 10 m above ground level should be averaged over at least 10 minutes and preferably over 15–20 minutes.

For all practical purposes, the relationship between fire spread and wind speed could be considered a straight line. However, Figure 4.10 illustrates the following important changes that occur in fire behaviour as wind speed increases.

- At zero wind speed, the rate of spread is equivalent to the backing rate of spread.
- Experimental field data suggest that, in fuel loads averaging 3–4 t/ha, a wind speed greater than 5 km/h at 10 m is required to consistently drive the fire forward as a heading fire. We have designated this the threshold wind speed.
- At wind speeds greater than the threshold (5 km/h in Fig. 4.10), the increase in rate of spread slowly lessens with increasing wind speed.

In very light winds – it is practically impossible to find dead calm conditions during the day – the thermal activity behind the fire front, from both residual burning material and the heated ground, draws the convective centre behind the leading edge of the fire. This convection can overcome the influence of the wind and draw the flames toward the burnt area, creating a backing fire all around the fire perimeter.

Light winds below the threshold wind speed are highly variable in direction and dominated by turbulent eddies, up-draughts and down-draughts. Under these conditions, fires are not driven in a consistent direction and spread alternately as heading fires as a result of down-draughts and backing fires when up-draughts occur over the fire area. Localised rates of spread may be as high as the value at the threshold wind speed but, because of the variability of direction and periods of backing spread, fires are easily controlled.

At wind speeds greater than the threshold, i.e. above 5 km/h, fires spread as heading fires in the direction of the prevailing wind. Well-documented experimental and wildfires suggest that the increase in rate of spread is slightly less than would result from a linear relationship with wind speed (Fig. 4.10). We could find no reliable evidence to support reports of a sharp decrease in the rate of spread at wind speeds above 50 km/h, and the effects of very high wind speeds (>80 km/h) on fire spread are uncertain.

The threshold wind speed for forward spread in discontinuous pastures is greater than 5 km/h. The actual threshold value depends on the fraction of bare ground in the pasture and the size of the patch of fuel (see 'Fire behaviour in spinifex' below). In heavily grazed and eaten-out pastures the threshold wind speed may be greater than 25 km/h.

We have considerable reservations about reports of head fires being blown out at high wind speeds. It seems possible that sections of a flank fire may be blown out when they are burning as backing fires in sparse fuels, and this may progressively narrow the head fire into long fingers. Eyewitness accounts seem to support this idea, but they differ on whether head fires can be directly extinguished.

During the 1978 Cyclone Alby fires in Western Australia, wind speeds exceeded 70 km/h with gusts recorded up to 150 km/h. One eyewitness described flank fires being blown out in sparse pastures, with the fire progressing across the paddock due to blown smouldering material, mostly sheep droppings. He considered the paddock would not have burnt at lower wind speeds. A second eyewitness expressed the opposite view, describing long fingers of fire in sparse fuels being blown out. He felt the fuels would have burnt at lower wind speeds.

During the Ash Wednesday fires in South Australia, head fires petered out as long fingers in sparse pastures when wind speeds were 45–50 km/h. After the wind change brought wind speeds of 70 km/h, reignition on the flanks of these fingers produced narrow fires that burnt across the sparse paddocks. This suggests that high winds were necessary to maintain a heading fire in the very sparse fuels.

Slope

While we have not been able to validate McArthur's relationship between forward spread and slope, relationships developed elsewhere in the world are in close agreement with it. We consider Figure 4.11 to be a useful guide to changes in forward spread up and down slopes.

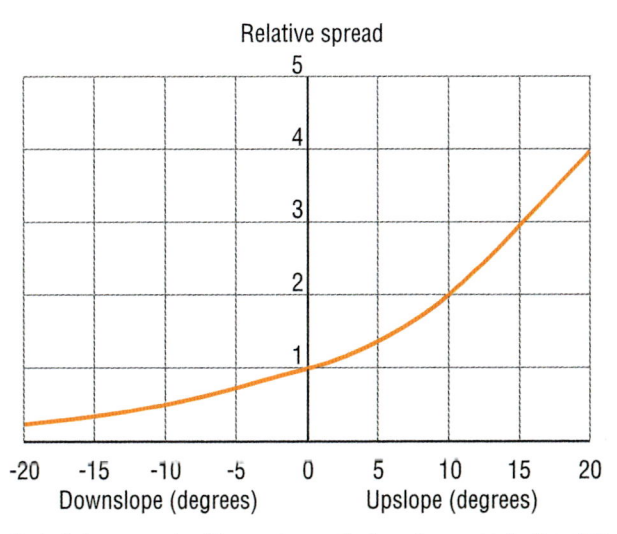

Figure 4.11: The effect of slope on rate of forward spread where 1 represents the relative rate of spread on level ground. Source: McArthur (1967).

A fire burning up a slope of 20° will burn four times faster than a similar fire on level ground. Wind speeds are generally higher towards the top of a slope than at the bottom, and this may produce further increases in the upslope rate of spread.

There is less certainty about the values of the multipliers when fires are travelling down-slope. If there is streamwise flow (see Fig. 6.10) and the wind is blowing parallel to the downslope surface the rate of spread may be similar to spread on level ground. However, a fire cresting a ridge usually creates an eddy wind on the lee slope, which complicates fire behaviour and can dramatically reduce rate of spread.

The direction of the prevailing wind is the dominant factor determining the direction of a fire, and strong wind will drive a fire across steep slopes. The whole question of the interaction of wind and slope is complex and is discussed in greater detail in Chapter 6.

Fire behaviour in spinifex

Hummock grasslands (see Chapter 2) occupy huge areas of the arid zone of Western Australia, the Northern Territory and South Australia. Comprised mainly of spinifex (*Triodia* spp. and *Plectrachne* spp.), their economic value has been regarded as low and in the past, under favourable seasonal conditions, fires have burnt unchecked over millions of hectares. Fire management is now recognised as important for pasture management on pastoral leases and to provide a diversity of plant ages for the conservation of native flora and fauna.

The most significant factors influencing fire spread in spinifex fuels are:

- fuel distribution;
- hummock size;
- fuel moisture content;
- wind speed.

Spinifex is a perennial grass, with hummocks expanding in diameter as the plant matures. The centre of a hummock will often die, leaving a living annulus or ring. As plants grow older these rings will often fragment, forming new individual clumps. The ground between hummocks is often bare.

The size and distribution of spinifex hummocks vary according to the maturity of the plants and the nature of the soil and terrain in which they are growing. Coverage of the ground by hummocks may vary widely, from near 100% in some drainage lines to less than 20% on rocky hills.

Because of the scattered and clumping nature of spinifex fuels, the spread and subsequent behaviour of fire in spinifex is very different from that in continuous fuels. For fires to spread, flames generated in one hummock must be able to reach the next. The length of flames will be determined largely by the hummocks' composition, size and moisture content (Fig. 4.12). The size of the gap between adjacent

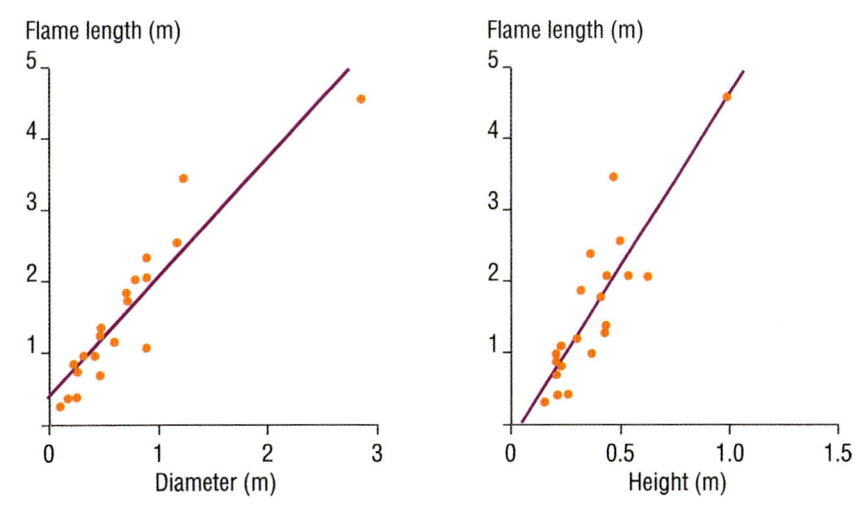

Figure 4.12: Relationship between maximum length of flame and hummock size (diameter and height) in Triodia irritans. Source: Bradstock and Gill (1993).

hummocks and the speed of the wind – the higher the wind speed, the lower the flame angle – will determine the length of flame needed for the fire to spread. Hence, ground cover and wind speed are the two most important factors affecting fire behaviour.

Fuel distribution

With the spatial continuity of the fuel strongly affecting fire behaviour – fire spreads more rapidly through a continuous fuel bed than through one that is patchy – various techniques have been used to describe a fuel bed's continuity. One expression of fuel distribution is the ratio of spinifex cover to bare ground; this has been used to predict fire spread in Western Australia's Gibson Desert. In general, it is sufficient to assume a relatively uniform scatter of hummocks and discuss fuel distribution in terms of spinifex cover, with the fuel load related directly to the fraction of spinifex cover.

Fuel moisture content

Both fine dead and fine live fuel moisture content are significant when dealing with most perennial hummock species, including spinifex. The first is influenced mainly by ambient air temperature and relative humidity, and the second by the moisture content of the soil, which is influenced by rainfall and evaporation.

However, it is usually not practicable to sort the live and dead spinifex for fuel moisture determination; fuel moisture contents have been measured by sampling the whole clump including both live and dead material. Under normal daytime conditions in the desert, fuel moisture has a relatively small effect on rate of spread

Figure 4.13: A spinifex fire in the Gibson Desert, WA, spreading under a strong wind which enables the flames to bridge the inter-hummock spaces. Photo: N. Burrows.

compared with other variables. Spinifex clumps are difficult to burn when their moisture content exceeds 35%.

Threshold wind speed

Spinifex hummock grasslands in the Gibson Desert, with 51–57% bare ground, require a threshold wind speed at 10 m of 15–21 km/h before heading fires will spread. They will not spread at all when the wind falls below the threshold, unlike the situation in continuous pastures. Similar work in Uluru National Park, where the hummock cover is greater, found that wind speeds >5–10 km/h at 10 m are required for fire spread.

When the average speed of the wind reaches the threshold value, the fire starts and stops erratically as the wind speed fluctuates above and below the threshold. The correlation between wind speed and fire spread is poor until the average wind is around 5 km/h greater than the threshold value and the fire spreads forward continuously. At higher wind speeds the rate of spread increases almost linearly with wind speed, parallelling fire behaviour in sparse and eaten-out pastures (Fig. 4.14). The wider the gaps between hummocks the higher is the threshold wind speed. Increases in the rate of spread above this threshold, while essentially linear, are at lower values than those in denser grasslands.

Because fire is spread by wind-driven flame contact, there may be little or no effective spread except in the direction of the wind. Often no spread occurs at the rear of the fire and there is little lateral spread on the flanks. As a result, spinifex fires are typically long and narrow, and they are often fragmented into long fingers.

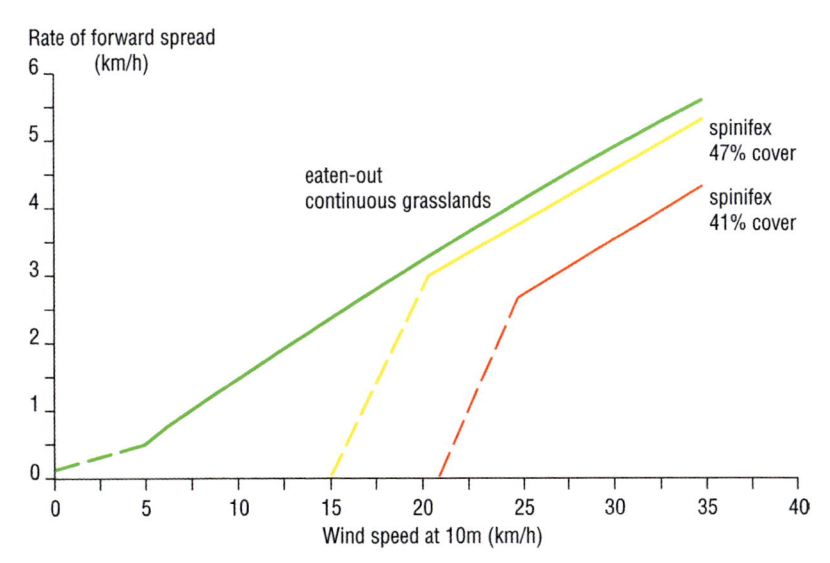

Figure 4.14: Rate of forward spread in spinifex for two degrees of fuel cover compared with spread in eaten-out grasses. The threshold wind speed must be exceeded before spinifex fires will spread. Source: Adapted from Burrows *et al.* (1991).

Lateral spread may occur in localised areas along the flanks where the hummocks are closer together; the fire can then increase its overall width when narrow head fires, originating from these clumps, burn along the flank. A substantial change in wind direction may result in numerous head fires developing where portions of the flank have remained alight.

Prescriptions for patch burning to minimise the size of wildfires and increase diversity in spinifex habitats need to specify the threshold wind speed for the fuel in question. Fires are lit in the morning after the wind speed increases above the threshold, and will spread until the wind speed drops below the threshold in the evening. The size of the patch likely to be burnt can be estimated from the length of time that the wind speed remains above the threshold and the mean wind speed.

Rainfall

As with most desert species, rainfall – and the lack of it – plays a big part in the life cycle of spinifex. Rainfall influences fuel moisture, as mentioned above, and fuel quantity. During periods of drought, spinifex hummocks cast off a considerable amount of foliage and become sparse, maintaining very low bulk densities. Fire behaviour is subdued under such conditions; flames are small and struggle to bridge the gap between hummocks.

Good rainfall may bring significant growth of short-lived grasses or other ephemerals between the hummocks. These will increase a fire's rate of spread, which may have characteristics similar to those in continuous pastures. Some spinifex fuels have been observed to burn twice within three years. Even after the short-lived grasses have been removed and the inter-hummock spaces are bare, new foliage growth thickens the hummocks and doubles the fuel load. Flames from individual hummocks are significantly longer than at other times, and threshold wind speeds are decreased. Fires in these fuel conditions may spread 2–3 km/h faster than predicted for drought conditions.

The regenerative response of seedlings or vegetative shoots from underground tillers, the severity of the fire and the general climate of the locality determine how quickly spinifex grassland fuels build up again after fire. In the absence of good rainfall and the growth of ephemeral species between hummocks, 15–20 years may be required in most arid regions before there is sufficient hummock grass fuel to carry another fire.

5
PREDICTING FIRE SPREAD

Chapter 4 showed how fires respond to changes in the key parameters – fuel, fuel moisture, wind speed and slope. Here we discuss how these factors are incorporated in the CSIRO grassland fire spread prediction system.

All fire spread prediction systems provide a singular result, which should be defined by the system's compilers. It is most important to know exactly what is being predicted and how that prediction has been obtained. The CSIRO system predicts the average potential maximum rate of forward spread of the head of a fire burning in uniform and continuous grassy fuel. A fire reaches its potential rate of spread after it has completed its build-up or development phase and its width is not constrained. The average, or quasi-steady, rate of spread should be taken as that measured over a period of 15–20 minutes; this accounts for short-term variations in spread caused by gusts and lulls in the wind and spatial variations in the fuel.

It is important to know exactly how the variables used to predict fire behaviour have been measured. Users must measure these in the way defined by the compilers to obtain the best results from the system. The CSIRO system uses standard meteorological observations of weather variables.

The most important factors that influence fire spread in continuous fuels are:

- fuel condition;
- grass curing state;
- dead fuel moisture content;
- wind speed.

Some factors that have a small influence on fire behaviour (e.g. seasonal condition and fuel factors) are built into the system by way of definition, while others that also have a small effect (e.g. rainfall and solar radiation) have been excluded. The following sections discuss how important factors are measured and their influence on fire spread.

The predictions are not applicable to discontinuous fuels such as spinifex grasses or the buttongrass moorlands of Tasmania – special guides should be used in these fuel types. The prediction system does not take into account the growth phase of a fire, and so will overestimate spread early in the development of the fire or if the head fire remains narrow.

Fuel condition

The predictions discussed below relate to fires burning in open country, woodland or open forest where the fuel bed is comprised of continuous annual or perennial grasses.

We have defined three distinct fuel conditions in open grasslands:

- undisturbed and/or very lightly grazed natural grassland or improved pasture, generally >50 cm tall (Fig. 5.1);

Figure 5.1: Ungrazed improved pasture.

Figure 5.2: Grazed pasture typical of many pastures in southern Australia in summer.

- grazed or mown pasture, generally <10 cm tall (Fig. 5.2). This is the common condition through the agricultural and pastoral zones of southern Australia, and is the pasture condition recommended for predicting the spread of fire across the landscape;
- very heavily grazed and eaten-out pasture, generally <3 cm tall, with scattered patches of bare ground (Fig. 5.3). This condition may be common in southern Australia during severe drought.

Fires in closely grazed or mown pastures will spread about 20% slower than those in natural/ungrazed grasses, provided both pasture types are continuous. We found that fires spread no faster in tall annual sorghums than in shorter kangaroo grass (*Themeda* spp.) pastures or natural grasslands of kerosene grass (*Eriachne burkittii*) flattened by receding floodwaters. Theoretically, rate of fire spread may be related to the height of the fuel bed but, as described earlier, there are practical difficulties in defining the height of the fuel bed.

In southern Australia, where most economically damaging grassfires occur, grasslands are generally grazed and are considerably shorter and more compact than ungrazed pastures. The relationship for fire spread derived from experimental data in mown grasslands in northern Australia, when extrapolated, is consistent with wildfire data gathered in southern Australia. We conclude that there is no

Figure 5.3: Eaten-out paddock. Fires in the middle ground have been unable to spread continuously under extreme conditions.

significant difference between grassfires in the two areas that cannot be explained by known variables.

In practice, a wildfire may burn over a variety of pasture types – including cereal crops and stubble fields, which may behave as either ungrazed or closely mown pastures – and rates of spread may vary accordingly from the values predicted obtained using only one pasture type. We recommend, however, that predictions of fire spread in southern Australia be based on relationships for closely mown or cropped continuous pasture unless one of the other pasture conditions clearly dominates the landscape.

For northern Australia, we recommend that predictions of fire spread in open grassland, grassy woodland and open forest be based on the natural pasture condition. Differences in spread in these vegetation types are determined by differences in wind speed near the ground.

Grass curing state

Most grasses have an annual or seasonal growth cycle, with senescence occurring following flowering (Fig. 2.8). 'Grass curing' is the term used to describe this process of dying and drying out. The curing state of a grassland, expressed as the fraction of dead material in the sward, has an important influence on both the ability of a fire to spread across the landscape and its rate of spread (Fig. 4.7).

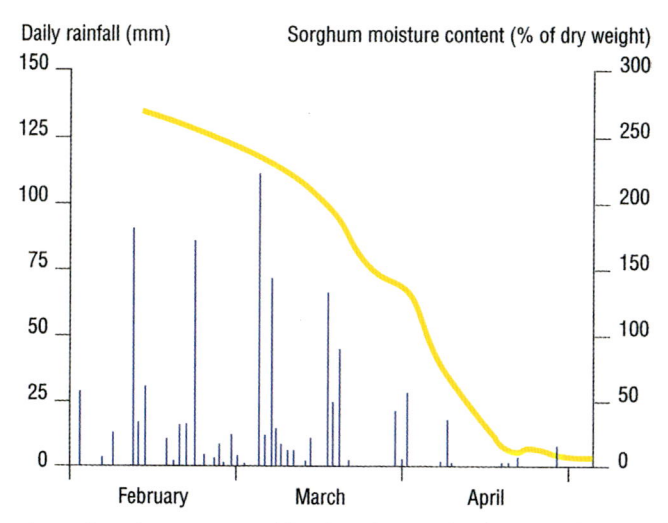

Figure 5.4: The change in moisture content of Sorghum intrans after flowering, Darwin, NT. Sorghum, an annual grass, does not reshoot after flowering and progressively loses moisture over the next three months despite frequent rainfall.

Grass curing state changes only relatively slowly. Annual grasslands commence curing once the grasses have flowered and set seed; the plants lose moisture over 6–10 weeks (Fig. 5.4). Once started, the curing process in annual pastures is not affected to any great extent by subsequent rainfall, although if the rainfall is sufficient to germinate seed green shoots may appear beneath the old sward. Such shoots are often not apparent at first (except where the old pasture has been burnt) and have no measurable effect on the rate of spread of fires in the fully cured older grasses that form a continuous sward. Once green shoots form more than 15% of the sward, however, they must be taken into account in the curing estimate and they will affect ignition and spread of fire. If rainfall is insufficient to maintain their growth, the green shoots will die before their life cycle is completed and they will have little effect on the overall degree of curing.

Perennial pastures cure more slowly than annual grasses, and curing is further delayed by rains early in the dry season. As perennials do not need to produce seed to continue their life cycle, rainfall after curing has started will delay the curing process in older leaves and produce new green shoots from the base of the clump that will continue to grow. Figure 5.5 shows typical patterns of grass curing for annual and perennial grasslands.

The rate of curing varies with soil type, soil moisture and daily evaporation. In general, annual grasses will become fully cured about six weeks after they begin to show distinct signs of yellowing (moisture content around 200%). This period is largely unaffected by intermittent rainfall but can be rapidly accelerated, perhaps by as much as a week, by a single day of strong, hot dry winds. Perennial grasses

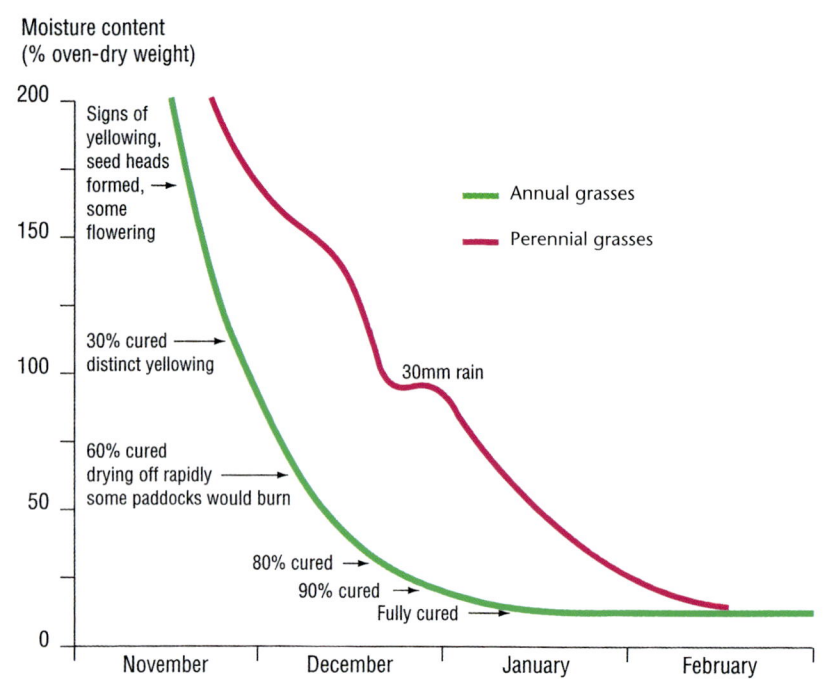

Figure 5.5: Curing of annual and perennial grasslands in the ACT during the summer of 1964–65. Source: Luke and McArthur (1978).

cure in a more irregular fashion. In coastal areas, with intermittent rain and high humidities, they may remain green all summer. If these grasslands have not been subjected to frequent grazing, however, the perennial grasses may carry a heavy sward of old dead growth from preceding years. Great care should be taken in estimating the fraction of dead material in the complete grass sward, particularly when the current growth looks fresh and green but may hide dead material underneath.

With practice, it is possible to make a reasonable visual estimate of curing state. However, because of spatial differences in curing rates, we should not expect an accurate prediction of fire spread across a particular landscape until all the grass is fully cured. Grass on dry ridges cures more rapidly than that in moist low-lying areas or creek lines. Therefore, early in the dry season fires that spread rapidly through fully cured pastures on ridges will be slowed by green grasses in depressions and stopped where pastures are less than 50% cured.

Satellite imagery used to estimate grass curing in southern Australia provides an integrated value of the curing condition of the landscape. A curing value of 80% for a 1 km^2 area may represent 100% curing over 85–90% of the area and less than 50% curing over the other 10–15%. Narrow bands of green grass along watercourses can provide an effective barrier to fires, and the rate of spread across a landscape assessed as 90% cured may, on average, be somewhat slower than across a continuous 90% cured pasture.

Difficulties have been experienced in using satellite imagery to estimate grass curing. To employ this technique, a reference image from the start of the season showing fully fresh green material is required. Subsequent images taken during the season are then compared with the reference image, and estimates made of the degree of colour change towards a fully cured condition. Estimates of grass curing derived from a reference image of one particular pasture may not be applicable to another. For example, tussock grasslands are rarely free of dead material, which makes it difficult to obtain a good reference image, and the relationship between colour change and curing is different from that for annual pastures. If there is a dry winter that results in little spring growth, it may be impossible to get a suitable reference image showing fully fresh green material.

Remote sensing techniques may be useful for some purposes, but it is essential that direct observation be made regularly during the fire season, both on the ground and from the air. On the ground, look for the nature of the curing, the difference between perennial and annual grasses, the amount of the previous season's dead growth under new green growth and the amount of new green material under recently cured growth – this condition commonly occurs during wet summers. From the air, get an appreciation of the variation across the landscape and the extent of uncured areas.

Once the landscape is more than 90% cured there is potential for widespread devastating grassfires. By this point there are few natural barriers, such as green creek lines and gullies, to inhibit the spread of fire. It is vital that frequent assessment of grass curing be made during fire season, particularly once grass curing has reached 70%.

Good data for deriving a relationship between curing state and rate of fire spread are also hard to obtain because of the large spatial variation in curing. Nevertheless, our experience suggests that the function illustrated in Figure 4.7 adequately describes the effect of curing on fire spread in uniform pastures.

Dead fuel moisture content

Dead fuel moisture content and wind speed are the two most important variables affecting how a fire spreads. Unfortunately, dead fuel moisture content is not easy to measure directly or to estimate in the field. It is measured experimentally by oven-drying samples of standing grass for 24 hours at 103°C.

Visual signs provide some guide to fuel moisture content, e.g. grasses are limp when they are moist and leaves become brittle and curl up when they are very dry. However, there is no reliable way of visually estimating moisture content in the relatively narrow range of 2–10% oven-dry weight where a change of 2% or 3% can substantially alter a fire's rate of spread.

Dead fuel moisture content can be estimated from the relative humidity and temperature of the air (see Fig. 6.13), but it does not appear as an independent variable in some prediction systems (e.g. McArthur Mk 4 grassland fire danger meter). Instead, the relationship between temperature, relative humidity and dead fuel moisture content is built directly into the equations or slide rules used for predicting fire spread or fire danger.

While the temperature of the air can be estimated roughly to within 5°, similar estimates of relative humidity are not readily made, so measurements are necessary. The temperature and relative humidity of the ambient air should be measured at a standard height of 1.2 m above the ground with wet and dry bulb thermometers in a Stevenson screen, or by using an aspirated psychrometer. Reliable electronic psychrometers are expensive and care should be taken to ensure they are properly calibrated and that the sensor is shaded from direct sunlight.

Rain or dew quickly wets grassy fuel to the point where it will not burn. However, once rain has stopped the dead fuel moisture content of the grass rapidly comes back into equilibrium with atmospheric humidity. On bright sunny days in summer, grassy fuel will dry out to equilibrium moisture content within two hours of quite heavy showers (Fig. 5.6).

Although rainfall obviously puts out fires, it is not included as a factor in predicting grassfire spread because its effect on dead fuel moisture content is short-lived. Firefighters should be aware that dangerous fire behaviour may occur soon after summer showers if the day is hot and dry.

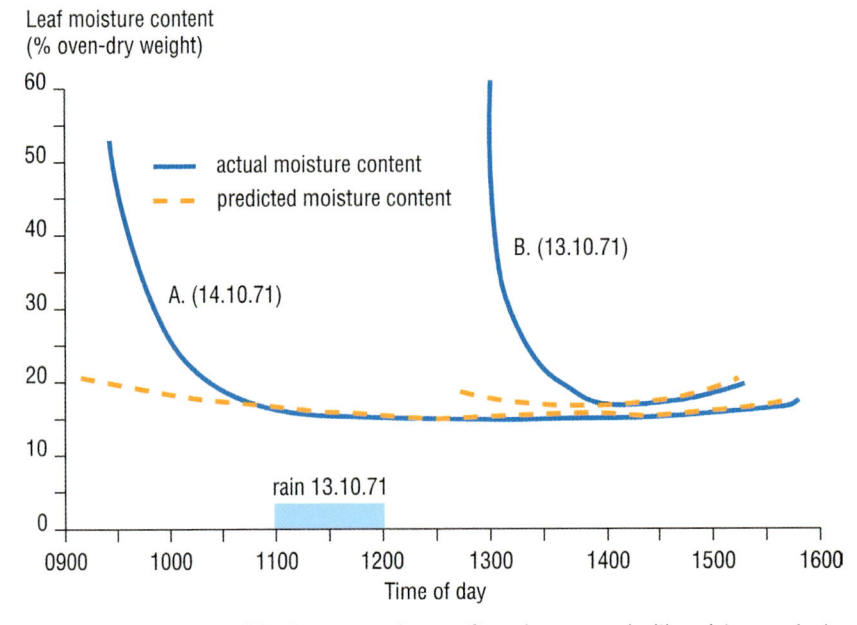

Figure 5.6: Moisture content of dead sugarcane leaves after rain compared with moisture content predicted from ambient temperature and relative humidity. A. Moisture content after 100 mm of rain overnight. B. Moisture content after 14 mm of rain between 11 a.m. and 12 p.m. that day. Note that sugarcane leaves are quite thick; fine grasses will dry out faster. Source: Cheney and Just (1974).

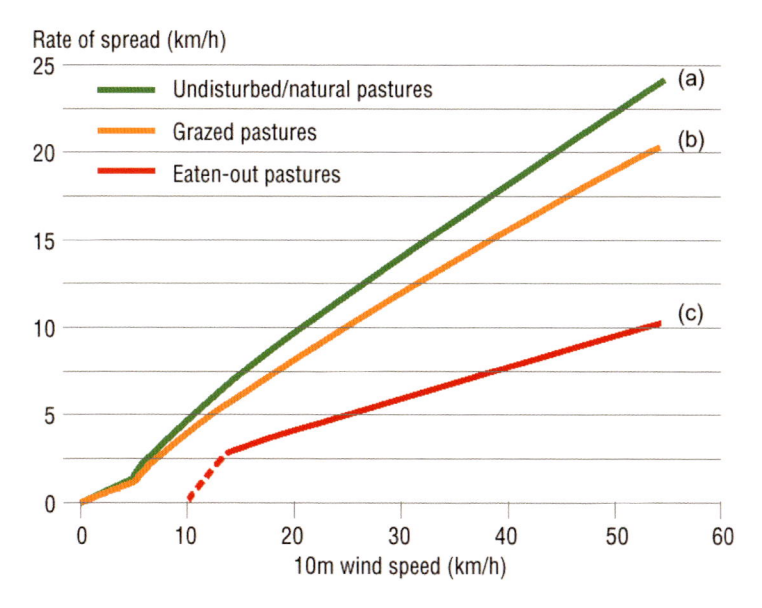

Figure 5.7: Relationship between average wind speed at 10 m and rate of forward spread for (a) undisturbed/natural grass, (b) closely cropped or grazed pasture and (c) eaten-out discontinuous pasture with a threshold wind speed of 10 km/h. During the day, when wind speeds are <5 km/h, we recommend that the rate of spread for a wind speed of 5 km/h be used.

Wind speed

All fire spread prediction systems used in Australia, including the one described in this book, are based on average wind speed. On occasions, it is suggested that the speed of the maximum gusts would better represent the fire danger. Fires respond, however, to both gusts and lulls, and both are incorporated in fire spread prediction through use of the average wind speed.

The standard measure of wind speed for fire spread prediction is the average value at 10 m above the ground in the open; it is recommended that this average be taken over 15–20 minutes. Measurements of wind speed at a height of 2 m in the open can be converted to the wind speed at 10 m by multiplying by 1.25. The relationship between fire spread and wind speed for the three classes of fuel condition are shown in Figure 5.7.

Provided grasslands are continuous, the rate of spread of fires is only 20% lower in closely cropped or grazed pastures than in tall undisturbed/natural grasslands. However, a fire in tall grass will have tall flames that may burn across narrow tracks, roads or firebreaks that would stop a fire with lower flame heights in grazed pastures. At lower wind speeds (<20 km/h), therefore, a fire's average rate of spread across the landscape is likely to be less than the potential rate of spread in continuous grasslands. When wind speeds are high, firebrands and embers can spread fire across quite wide firebreaks and the average spread across the landscape will approach the potential rate of spread in continuous fuels.

Dead calm conditions rarely occur during the day. When the wind speed at 10 m is <5 km/h, we recommend using the rate of spread in a 5 km/h wind because:

- wind near the ground is variable and, although the direction of the wind and therefore the direction of fire travel is erratic, there may be quite strong local puffs >5 km/h;
- the stall speed of many anemometers is around 3 km/h, so the average wind speed is likely to be greater than that recorded by the anemometer.

In severely eaten-out pastures the threshold wind speed required for continuous spread depends on the continuity of the fuel. This may vary from around 5 km/h to >20 km/h. We have used a threshold wind speed of 10 km/h in Figure 5.7 only for illustration. This feature of grassland fire behaviour may catch firefighters unaware, particularly if they are relying on the eaten-out pastures to contain a fire.

During the 2003 ACT fires the eaten-out pastures between Canberra suburbs and the mountain forests gave firefighters a false sense of security, leading some to believe that the fire could be held in those fuels even though extreme fire danger weather was forecast. On the morning of 18 January the fires burning in the forest west of Canberra had been checked along the eastern perimeter as it entered sparse grassland. Firefighters had little problem suppressing spot fires starting in the grass until the wind speed exceeded 25 km/h, after which the fires broke away and burnt across eaten-out pastures at 3–4 km/h. At 2 p.m. the weather conditions recorded at Tidbinbilla were temperature 37°C, relative humidity 8% and mean wind speed of 55 km/h. Between 2 p.m. and 2.30 p.m. the fire spread across sections of eaten-out and ungrazed pastures at an average speed of 11 km/h. Our predicted rates of spread were 7.5 km/h and 17.5 km/h respectively.

A wildfire in southern Australia is likely to encounter all three fuel conditions if it runs over a number of kilometres. Likewise, the wind speed at the fire site may be different from that at the anemometer station. The fire spread prediction system provides a useful guide to potential rate of spread, but users should expect some variation around the predicted value depending on the actual fuel and wind conditions at the fire.

The presence of trees reduces the wind speed near the ground and thus the rate of spread in woodlands and open forests with a grassy understorey. The ratios between wind speed at 10 m in the open and at 2 m in woodland and open forest and the rate of forward spread relative to forward spread in the open are given in Table 5.1. The rate of forward spread in woodland and open forest is less than would be obtained by simply applying the wind speed ratio. These rates of relative forward spread are incorporated in the CSIRO system for fire spread prediction in northern Australia.

In practice, the height and density of trees will vary across the landscape, from patches of open grassland through scattered woodland trees to open forest with

Table 5.1: Ratios between wind speed at 10 m in the open and 2 m above the ground and relative fire spread in different types of vegetation

Type of vegetation	Ratio between wind speed at 10 m in the open and at 2 m	Rate of forward spread relative to spread in the open
Open grasslands	10:8	1.0
Woodlands (5–7 m)	10:6	0.5
Open forests (10–15 m)	10:4.2	0.3

30% canopy. The strength of the wind near the ground will vary accordingly, as will the rate of fire spread. The relative rate of spread for woodlands (Table 5.1) should provide a reasonable estimate of fire spread through such a mixture of vegetation.

Slope

Predictions of fire spread are generally made for level to gently undulating topography where the differences in spread on positive and negative slopes will cancel each other. Corrections for fire spread up slopes can be made using the multipliers given in Chapter 4, and a number of fire spread models assume that corresponding reductions occur on negative slopes. In practice, the large number of calculations required to predict the spread of a fire through broken terrain requires the use of computer-based models (Fig. 5.8).

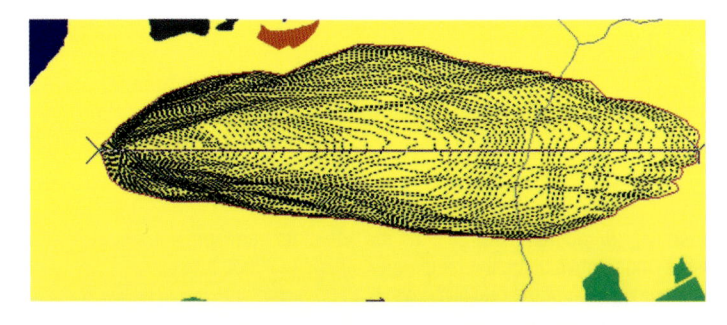

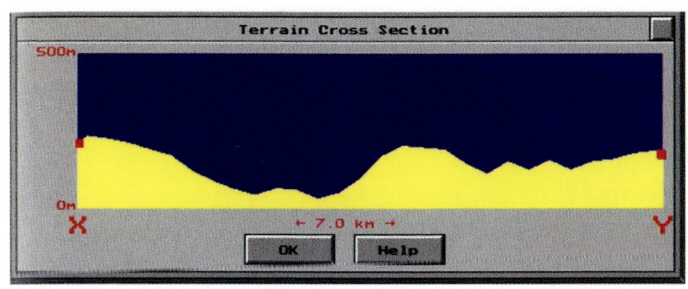

Figure 5.8: Computer simulation of the spread of a grassfire across hilly terrain using the CSIRO bushfire spread simulator SiroFire. The lower diagram shows a cross-section of the topography along the length of the long axis of the fire.

Steeply divided terrain influences wind speed and direction near the ground, and may create lee-slope eddies (see Fig. 6.11). A fire burning against a lee-slope eddy will travel at its backing rate of spread, and fire spread models that assume only the reduced rate of forward spread on negative slopes given in Figure 4.11 will overestimate the rate of spread.

CSIRO grassland fire spread prediction system

The mathematical equations describing the graphs in Chapter 4 form the basis of this fire spread prediction system. These equations (see the literature cited in the bibliography) can be expressed in other ways, such as nomograms or circular slide rules, that are useful for illustrating the relative importance of each variable and are easy to use in the field.

Fire spread nomogram

A nomogram is a series of interlinked charts that form a graphical aid for solving mathematical equations. Figure 5.9 shows a nomogram for predicting the rate of forward spread of fire in natural or ungrazed grasslands.

This nomogram can also be used to predict the rate of spread of fires in woodland and open forest in northern Australia. The rate of spread in woodland is about half that in open natural grassland; the rate of spread in open forest is about one-third of that in open natural grassland.

A similar nomogram to predict the rate of spread in grazed pasture is shown in Figure 5.10. The rate of spread in eaten-out pasture is approximately half the rate in grazed pasture.

Figure 5.9 (Opposite):   Nomogram for predicting the rate of forward spread for grassfires in natural/ungrazed pastures, using relative humidity, temperature, dead fuel moisture content, degree of curing, wind speed and slope in the direction of the prevailing wind.

The rate of forward spread is estimated by tracing a line through the charts to connect the values of the variables of relative humidity, temperature, dead fuel moisture content, degree of curing, wind speed and slope. For example, we may want to calculate the potential rate of fire spread in natural pastures under the following conditions:
- relative humidity 30%
- air temperature 30°C
- degree of curing 100%
- average wind speed at 10 m in the open 30 km/h
- slope in the direction of the wind +5°

1. Starting at Chart 1, trace a line horizontally from 30% relative humidity to the diagonal temperature line of 30°C.

Predicting fire spread | 61

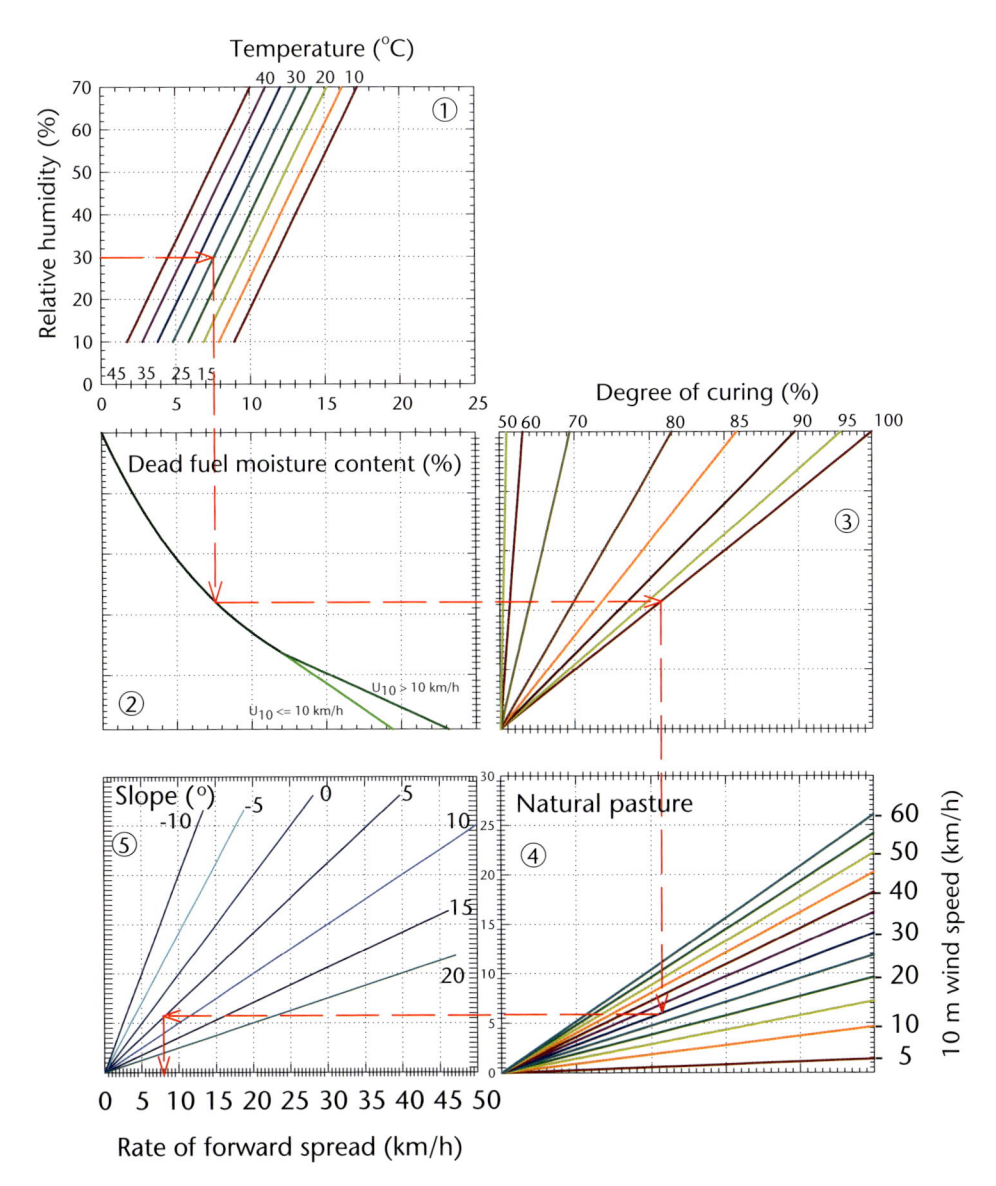

2. At the point of intersection, trace a line vertically downwards to the curved line in Chart 2. (The point where the vertical line intersects the X axis of Charts 1 and 2 gives an estimate of the dead fuel moisture content of the grass (7.8%). If dead fuel moisture content is known, fire spread can be calculated by starting with this value at the top of Chart 2.)

3. From the intersection of the vertical line and the curved line in Chart 2, trace a horizontal line to the 100% degree of curing diagonal line in Chart 3.

4. Now trace vertically downwards to Chart 4 to the diagonal line for an open wind speed of 30 km/h.

5. From this point, trace a horizontal line across to the 5° slope line in Chart 5.

6. At the intersection of the horizontal line and the slope line trace a line vertically downwards, and read off the rate of spread of 8 km/h at the bottom of Chart 5.

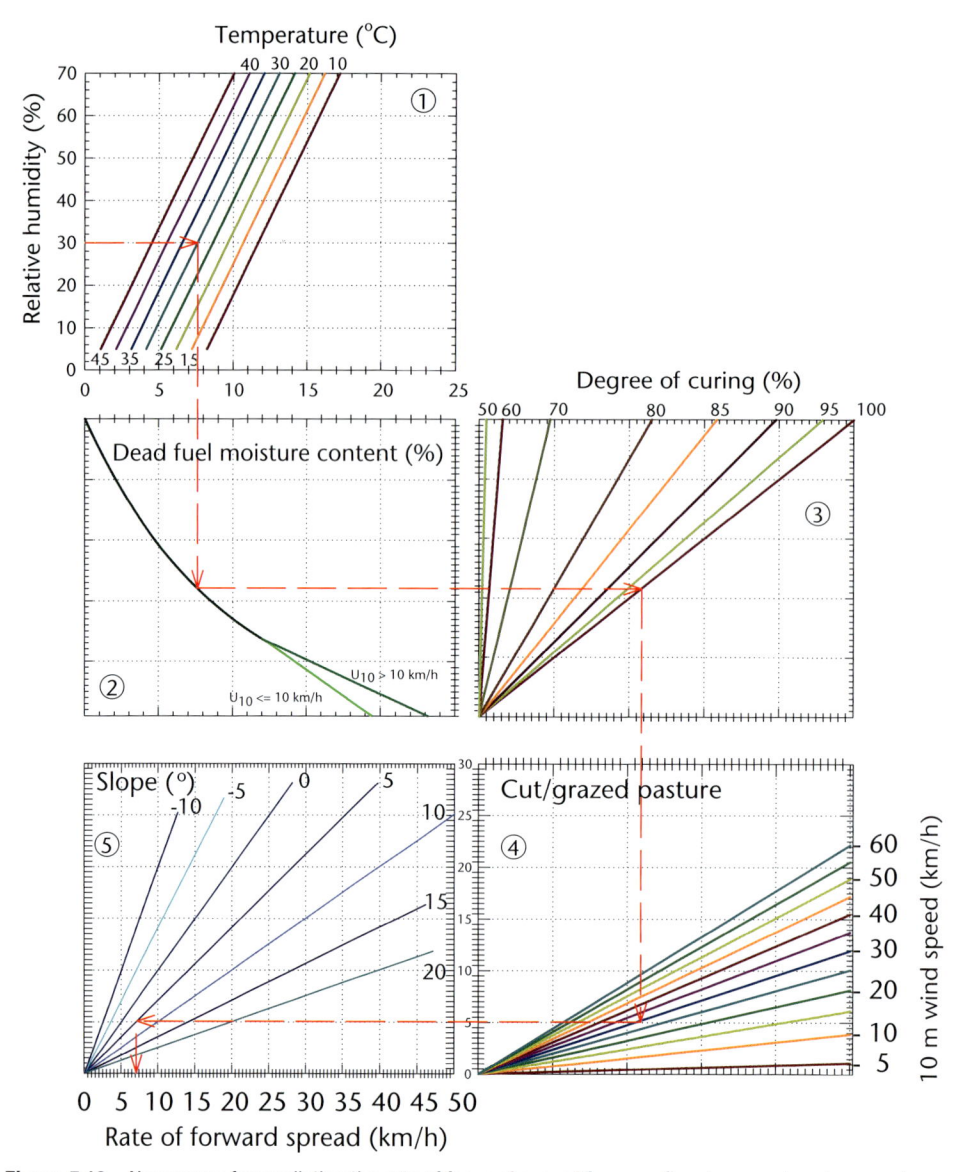

Figure 5.10 Nomogram for predicting the rate of forward spread for grassfires in grazed pastures, using relative humidity, temperature, dead fuel moisture content, degree of curing, wind speed and slope in the direction of the prevailing wind.

Fire spread meter

The same calculations can be performed using a circular slide rule. Figure 5.11 shows the CSIRO grassland fire spread meter, which predicts a fire's potential rate of forward spread across continuous grassland in undulating terrain. Slope is not

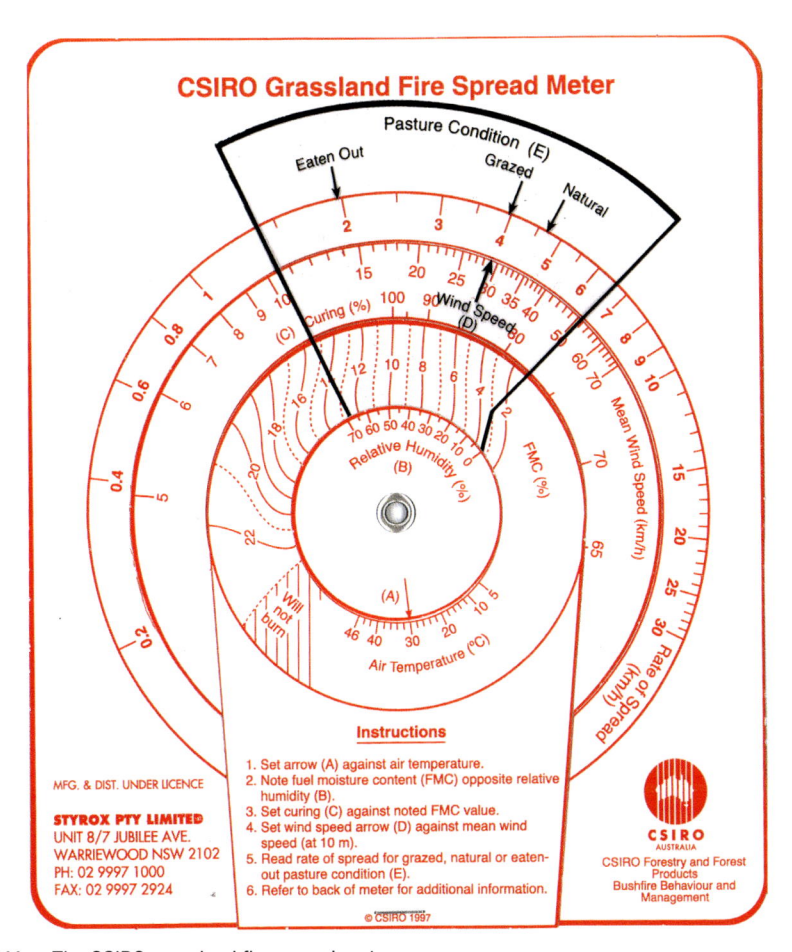

Figure 5.11: The CSIRO grassland fire spread meter.

included. The scale of the meter is not appropriate for determining rates of spread at zero wind speed. In any case, for practical purposes it is better to use a minimum wind speed of 5 km/h when winds are light and variable during the day.

Both the nomogram and the grassland fire spread meter predict the average potential rate of forward spread of the head fire over periods of 15–20 minutes. The short-term variation around this average value is quite high; users may find rates of spread in the field varying by ±25% of the predicted value if they make observations over periods of less than 10 minutes.

At high wind speeds, the potential rate of spread of a fire is not achieved until the width of the head fire is >200 m. Therefore, the meter will overpredict rates of spread in the first 15–20 minutes of a fire's development, or if the head fire remains narrow for any reason. The predicted potential rate of spread may be greater than the observed rate for wildfires burning under low to moderate wind speeds. This is

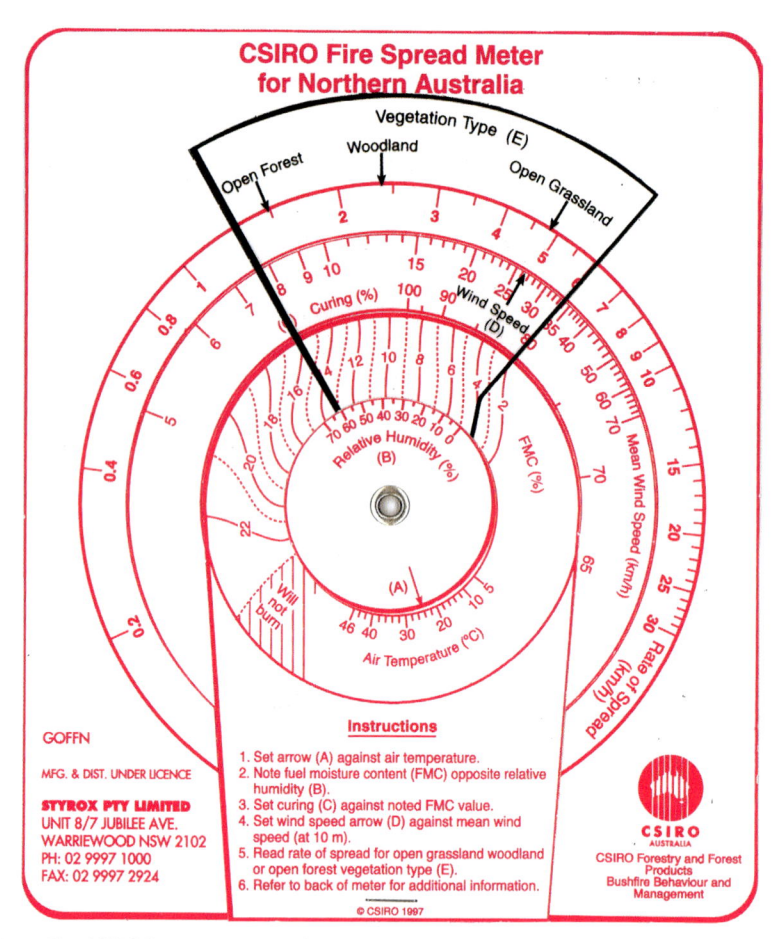

Figure 5.12: The CSIRO fire spread meter for northern Australia.

because, under these conditions, fires may be held up for brief periods by roads, green gullies, firebreaks etc. so that the average rate of spread determined by timing a fire's movement past features in the landscape may be substantially lower than the spread across continuous pasture.

This meter uses the equation for fire spread in grazed grass fuel and approximates spread in natural/ungrazed fuel by multiplying the grazed rate of spread by a factor of 1.2. Fire spread in eaten-out grass fuels is half that of the grazed rate of spread.

Fire spread meter for northern Australia

A fire spread meter for northern Australia has been designed to predict the rate of spread of fires in open grassland, woodland and open forest with a grassy understorey (Fig. 5.12). This meter is not suitable for predicting fire spread in tall closed forest or tall forest with a substantial shrub and litter component in the understorey. It is designed to predict fire spread where the grassy fuels are largely undisturbed or grazed only lightly.

This meter uses the equation for rate of spread in natural/ungrazed fuel with the relative rate of spread multipliers for woodland and open forest given in Table 5.1.

6
LOCAL VARIATION AND ERRATIC FIRE BEHAVIOUR

As described earlier, grassfires can appear highly erratic, with behaviour that is very variable and difficult to predict. Much of this variation can be taken into account by taking average values of fuel and wind conditions and predicting average rates of spread over periods of 15 minutes or more. However, it is important for firefighters to understand that, over much shorter periods, grassfires will respond to many factors that cannot be included in a fire spread prediction system and are, for all intents and purposes, unpredictable.

Not only does a fire respond to changes in fuel and weather in different ways but these factors themselves are rarely constant; they are highly variable in time and across the landscape.

Wind

As we have seen, wind plays an important role in determining the behaviour of a grassfire. Rapid changes in wind direction and strength can cause immediate changes in a grassfire's behaviour. Wind varies greatly in strength and direction, both in time and in location. Wind at one point in the landscape will be different at another time but it will also be different somewhere else at the same time – sometimes even quite near. This variation is a result of turbulence embedded in the wind flow that forms gusts (increases in wind speed over the average) and lulls (decreases in wind speed over the average). Meteorologists generally describe the surface wind in statistical terms (i.e. average wind speed and direction) and use values that are averaged over defined intervals, e.g. 10–20 minutes, in order to overcome the variations caused by the gusts and lulls. The turbulence generally takes the form of eddies, that result from a number of mechanisms.

Eddy formation

Wind near the ground is the result of a complex interaction between the atmosphere, the earth and the vegetation on the earth's surface. The rotation of the planet and the friction between the atmosphere and the earth's surface induces a fast-moving wind in the upper atmosphere, called the geostrophic wind. Heating of the earth's surface by the sun on hot days results in the formation of convective cells or thermals of rising air. These rising bubbles of air mix with the lower layers of the geostrophic wind. The height at which this mixing occurs defines the atmospheric boundary layer. During the night this layer is quite shallow, coming down to only a couple of hundred metres from the surface, but during the day when convective heating occurs the atmospheric boundary layer increases in height. It can be about 1.5 km thick during a typical summer day.

When the rising convective bubbles of air interact with the fast-moving geostrophic wind, rotating rolls called eddies form in the shear between the two layers (Fig. 6.1), similarly to the way a pencil will roll between your hands when you slide one hand against the other (Fig. 6.2). These eddies can range in size from the whole depth of the boundary layer down to a few millimetres, and exist simultaneously throughout the whole boundary layer. The eddies are constantly decaying, breaking up into smaller and smaller eddies, in what is known as an energy cascade, while new eddies are continuously formed. Wind gusts felt on the surface are the result of these eddies in the boundary layer striking the ground as they travel past.

Depending on the size of the eddy, the gust may be felt locally or over a wide area. During days when the atmosphere is unstable there may be a repeated general overturning of the boundary layer every 15–20 minutes (i.e. 3–4 times an hour), forming large boundary layer-sized eddies that result in a sudden gust over a kilometre or more as the fast-moving air of the lower geostrophic layer is brought down to the surface. These gusts can occur even on a relatively calm day and give a short-lived impression of a windy day.

The flow of wind across the earth's surface is further complicated by mechanical turbulence introduced by obstructions such as buildings and trees. These obstructions introduce wake eddies that are superimposed upon the eddies that exist in the general wind flow. Obstructions also act to slow the flow of these eddies and accelerate their break-up into smaller forms.

Local variation and erratic fire behaviour | 69

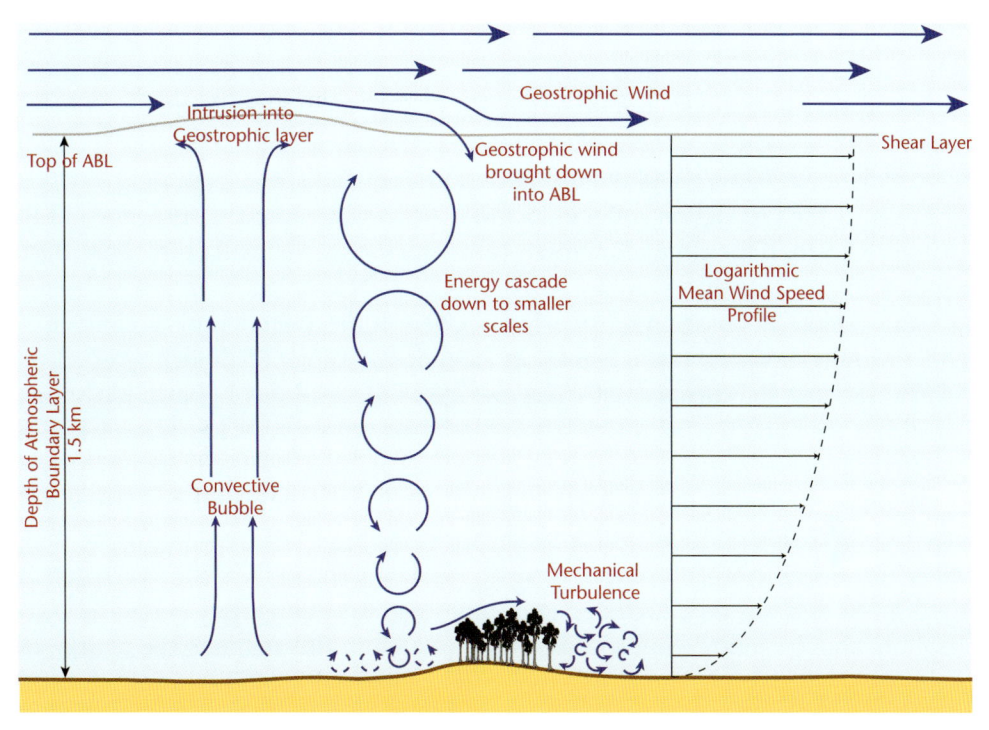

Figure 6.1: Formation of eddies in the atmospheric boundary layer, which causes the gust and lull structure of wind.

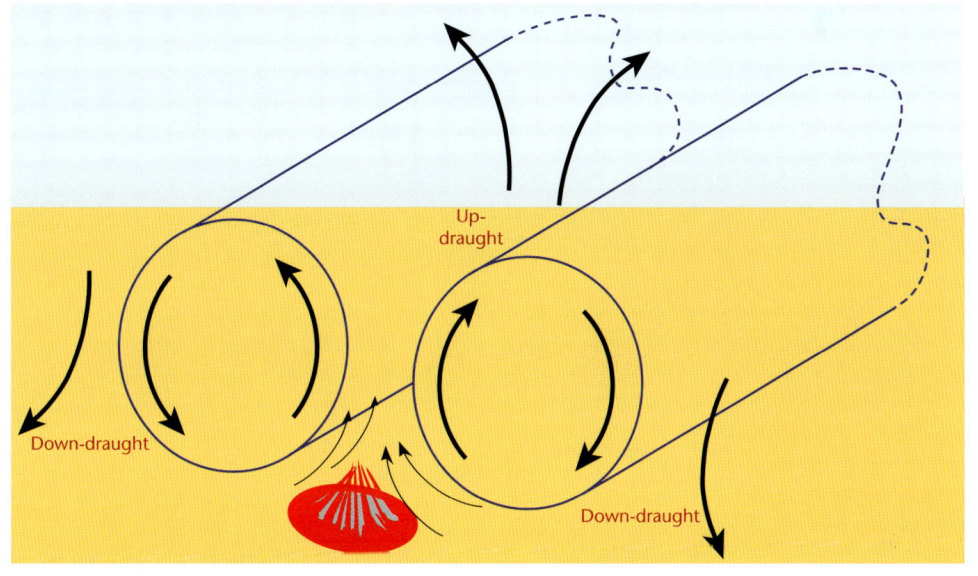

Figure 6.2: Eddy circulation in the atmospheric boundary layer and resulting wind directions on the ground under a light and variable prevailing wind.

Resulting wind variation

When eddies are carried along with the prevailing wind they cause winds at the surface to vary in strength and direction, depending on the speed and direction of rotation of the eddy. Figure 6.5 illustrates this effect. In this example, the speed of the circulation is 8 km/h and the prevailing surface wind speed from right to left is 20 km/h. If the circulation rotates clockwise (a) it will cause the surface wind to gust to 28 km/h. If the circulation rotates anti-clockwise (b) it will cause the surface wind to lull to 12 km/h.

An example of variation in the wind is given in the anemogram in Figure 6.3. This diagram shows how wind speed at Essendon Airport, Melbourne, varied on Ash Wednesday, 16 February 1983. The average wind speed was 37 km/h (20 knots) between 3 p.m. and 6 p.m., and gusts and lulls ranged from 81 km/h to 18 km/h. The sharp increase in wind speed at 8.45 p.m. was associated with a frontal change.

While an anemogram shows variation in wind speed with time at a particular site, it is a little more difficult to appreciate the patterns of similar but different variations that occur in other locations close by. Figure 6.4 shows the spatial variation of wind speed that was measured at one instant in a 60 × 80 m forest block and converted to above-canopy wind speeds. When plotted in this manner, the pattern of wind speed has the appearance of a topographic map, with a lull creating a 'valley' between two 'mountains' formed by gusts at the top and lower right of the block. This pattern is similar to patterns measured over open pasture to explain gust structure for sailors.

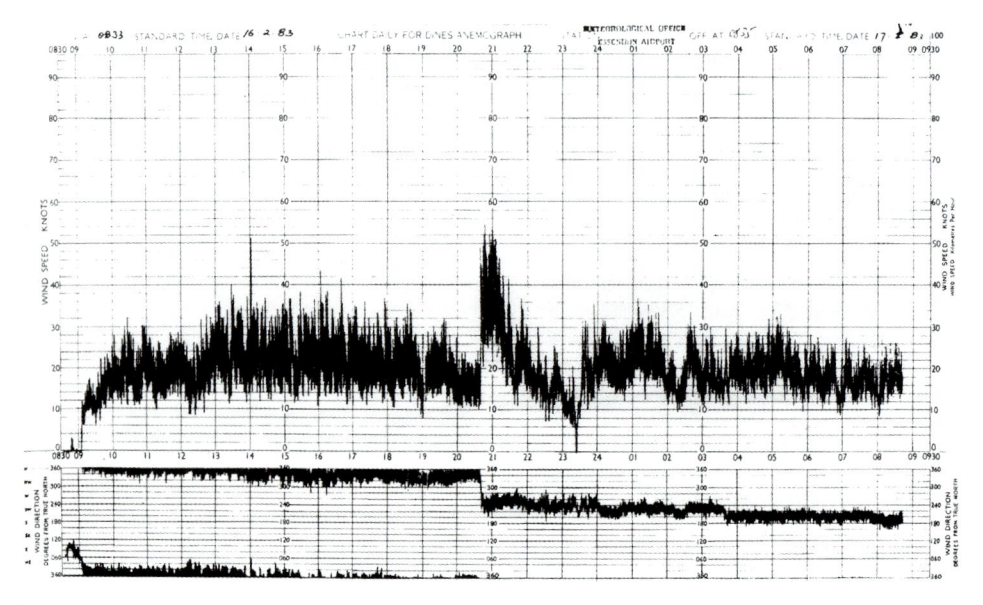

Figure 6.3: Anemogram from Essendon Airport on Ash Wednesday, 16 February 1983. Source: Bureau of Meteorology (1984).

Local variation and erratic fire behaviour | 71

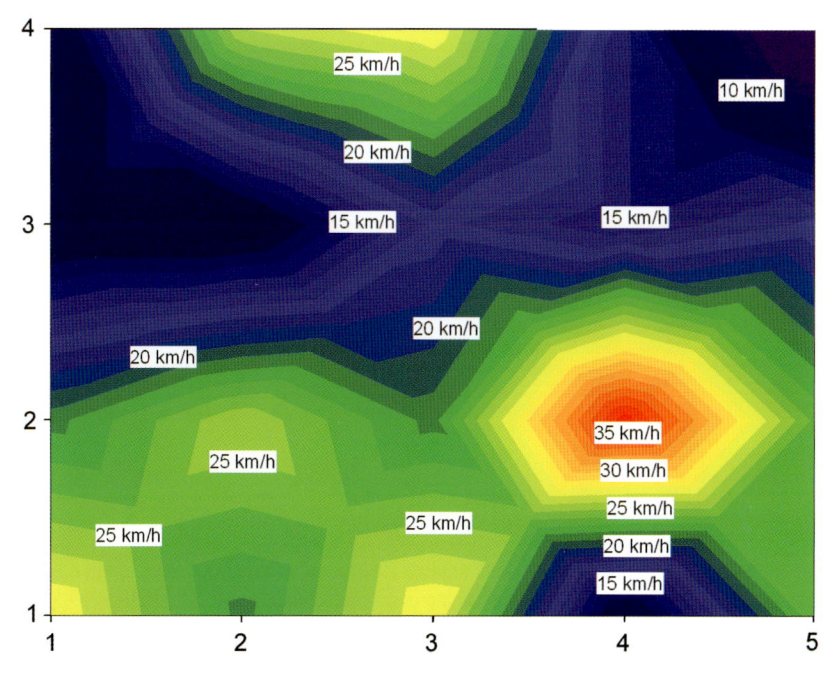

Figure 6.4: Spatial variation of wind speed (km/h) in a 60 × 80 m block at a single instant as measured by 20 anemometers laid out in grid of four rows and five columns, with 20 m between each row and column.

In this situation, a fire burning in the bottom right corner of the block would respond to the strong wind gust, while an anemometer on the edge of the block would record a much lower wind speed than that driving the fire front. This is why it is necessary to take wind measurements over a relatively long period – so the average wind at the anemometer will be close to the average wind driving the fire.

Because wind speed increases with height above the ground, due to the cascade to smaller eddies close to the ground, the spatial variation of the wind at any height will differ from that on the ground.

Fire response to gusts and lulls

Sudden changes in the speed and direction of a fire can hamper firefighters' efforts to bring it under control. Grassfires respond almost immediately to changes in wind speed and direction, so it is important that firefighters understand how and when wind changes may occur, a fire's reaction to these changes and how their personal safety may be affected.

The convective cells or thermals set up in the atmosphere on a hot day can result in local up-draughts beneath the convective cell, and down-draughts between cells. When a down-draught strikes the ground, the surface winds tend to spread in all directions. Conversely, surface winds tend to be drawn together to the base of the

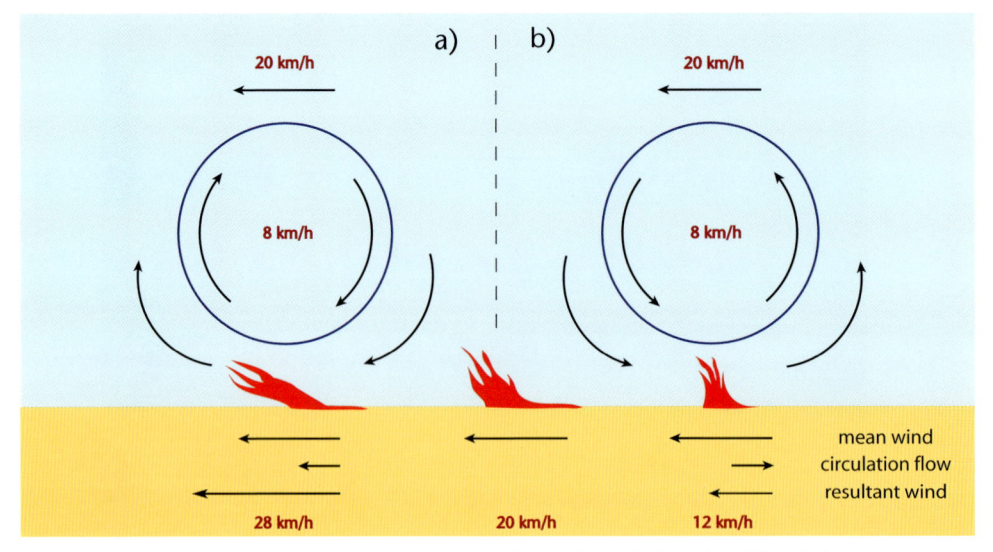

Figure 6.5: Eddy circulation superimposed on a prevailing surface wind moving right to left causes variability in the surface wind speed. (a) A fire burning under the influence of a clockwise rotating eddy will experience increased wind speed (a gust) and the flames will be flattened. (b) A fire burning under the influence of an anti-clockwise rotating eddy will experience decreased wind speed (a lull) and the flames will stand up.

up-draught. In general, the area affected by the down-draught is more extensive than that affected by the up-draught.

The flames of a fire burning in this wind field would generally be flattened by the down-draught associated with the gust (see Fig. 6.5a). During a lull, the flames would stand more upright (Fig. 6.5b). At times, the convection column of the whole fire may connect with the up-draught, enhancing the effect of the lull. The flames at the front of the fire would then be drawn back over the burnt area, so the fire would stall to its backing rate of spread.

When a down-draught strikes the surface it enhances the strength of the surface wind, but also causes sudden shifts in wind direction. Small fires can respond to these shifts quickly and, depending on their location, may travel in quite different directions (Fig. 6.6). Successive down-draughts that cause such a fire to change direction frequently during its initial period of spread will also cause it to build up rapidly, as illustrated in Figure 6.7.

Thermal up-draughts travelling over a fire can have dramatic effects on fire spread and behaviour. Figure 6.8 shows a pair of simultaneous fires burning in tall standing natural pasture (B244) and closely mown pasture (B241). Although the wind speeds at six locations on the block corners were similar, the winds affecting the fire fronts were quite different. Fire B244 was under the influence of a thermal up-draught. This produced a pointed head fire and both the flames and convection column were more erect, causing the fire to slow down.

Local variation and erratic fire behaviour | 73

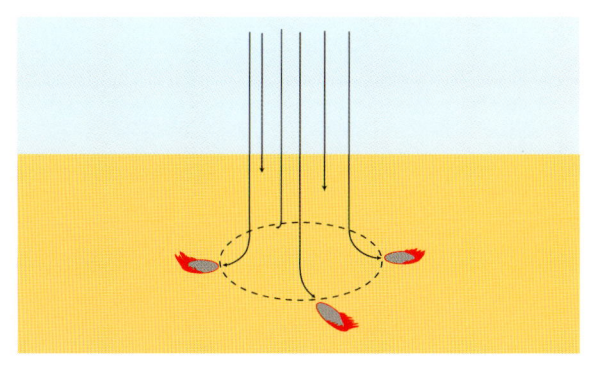

Figure 6.6: Effect of a strong down-draught striking the ground. When fires are small their initial spread may be in quite different directions.

Fire B241, burning in mown grass, was under the influence of a thermal down-draught. This enhanced the surface wind and caused the head fire to spread out in a parabolic shape, with both the flames and the smoke lying down at a low angle. As a result, this fire spread faster than fire B244 burning in undisturbed grass despite the fact that, under equivalent conditions, fires in undisturbed grass travel on average 20% faster than fires in mown pasture.

Fires spreading in woodland tend to show a greater response to thermal activity than fires in the open, because of the lower wind speeds below the canopy near the fuel. It is common for a fire in woodland to stop completely when a thermal up-draught is established over the front even though winds outside the fire may appear quite strong (Fig. 6.9).

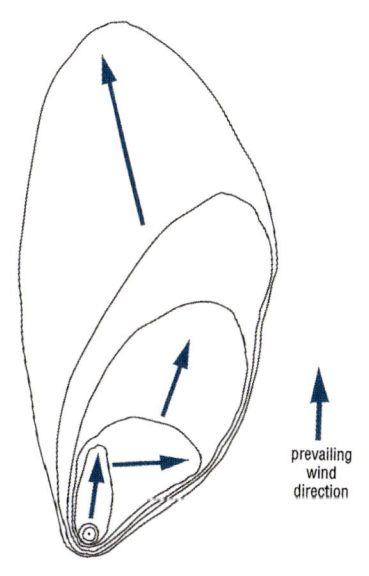

Figure 6.7: Fire 25 Gunn Point, NT, 1974, showing a rapid increase in rate of spread in response to frequent changes in wind direction during the first 14 minutes.

Figure 6.8: Experimental fires B241 (bottom) and B244 (top), NT, 18 August 1986, 2 minutes after simultaneous ignition. An up-draught over fire B244 is causing the smoke and flames to rise at a steep angle and the fire to slow. A down-draught over fire B241 is flattening the smoke and flames and causing the fire to speed up. B244 is burning in undisturbed natural pasture and would have burnt out of the plot by this time if the wind speed behind the two fires had been exactly the same.

As a fire becomes larger and more intense, its front will integrate some of the turbulence in the wind field. For example, experimental fires travelling at 5 km/h under a mean wind speed of 15 km/h in open grassland did not respond to variations in the wind speed of ±5 km/h that came at intervals of around 20 seconds. They did respond to greater variations coming at around 4 minute intervals.

Wind and the terrain

A hill – or any other topographic obstacle – will interfere with the flow of the wind field, changing the wind's speed, direction and turbulence. The mechanics of these changes are not well understood. Generally, wind speed will progressively increase up the slope to the crest of the hill. The wind from lower elevations will be compressed by the wind above it and forced to speed up. Figure 6.10 illustrates this compression of wind flow on the windward side of a hill, resulting in increased wind speed.

The wind flow on the lee side of hills or ridges is also a complex matter. Depending on the steepness or curvature of the ridge, the speed of air flow and the stability of the lower air layer, the air flow may be 'streamwise' or 'recurved'. Streamwise

Local variation and erratic fire behaviour | 75

Figure 6.9: Experimental fire 17 in open forest, Gunn Pt, NT, 17 August 1974. (a) At 14 minutes after ignition a down-draught has flattened the flames at the head of the fire and is keeping smoke close to the ground. (b) At 21.5 minutes after ignition an up-draught has enhanced the convection behind the front and spread has stalled.

Figure 6.10: The streamwise flow of wind over a hill. Wind speed near the surface progressively increases as the top of the hill is approached.

flow will result in the wind blowing down the lee side of the hill (Fig. 6.10). Under conditions causing recurved flow, the prevailing wind separates and lifts away from the hill, creating an eddy flow back up the hill in the direction opposite to the prevailing wind field (Fig. 6.11). It is not easy to predict whether the wind flow over a hill will be recurved or streamwise.

Very high rates of spread can occur on steep slopes, due to both the effect of the slope and the increase in wind speed on it. Fortunately, these rates of spread are short-lived. A grassfire will generally be slowed markedly on a lee slope; eddy-winds may reduce the progress of the fire's front to its backing rate of spread as it travels down the slope. Provided access is not a limiting factor, grassfires in divided country can be easier to suppress than those on level terrain where there is no impediment to fast-spreading head fires. However, where remnant trees remain on ridge-tops, spotting can cause additional suppression difficulties.

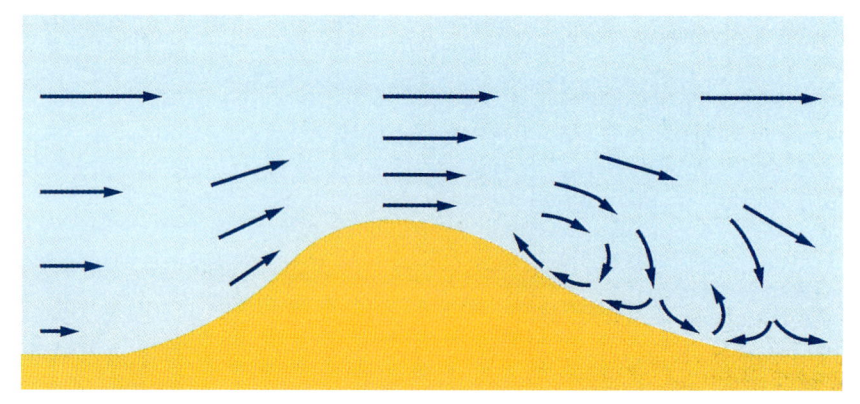

Figure 6.11: Separation of wind flowing over a hill forming a lee-slope eddy. A fire travelling over the hill will burn down the slope as a backing fire under the influence of the eddy. Under these conditions, rates of spread downslope will be less than illustrated in Figure 4.12.

The speed and direction of wind near the ground is very variable and difficult to predict in hilly country. Winds can sometimes flow at right angles to the direction of synoptic winds, mainly due to the channelling effect of steep gullies. In some locations within the topography, it may be quite difficult for firefighters to estimate the direction of the prevailing wind.

When winds are light, the slope will be the dominant influence on the direction of fire spread. The convection of the fire can enhance local upslope thermal winds, giving the impression that quite a strong wind is blowing the fire upslope. Under strong winds (>15 km/h), the wind direction will determine the direction of fire spread and fires may be driven across quite steep slopes. If there is a lull in the wind or the wind direction aligns with the slope, the flank of the fire will progress rapidly upslope.

Heating of the ground during the day and subsequent cooling at night also induce winds in hilly terrain; this effect is most noticeable at night but can also be seen during the day. During the day, gully winds almost always flow up the slope. These upslope (anabatic) winds are generally turbulent and stronger than winds on uniform slopes. The direction of flow is reversed at night. Katabatic (downslope) winds are denser, slower and less turbulent than upslope winds.

Another example of terrain affecting wind flow is the mountain or lee wave phenomenon. Strong winds that move across a mountain range can form large-scale eddy rolls on the lee side, and these will drag fast-moving air from the upper levels down to ground level some distance from the mountain range. These strong winds are extremely localised; conditions may vary from near gale force winds to calm within a few kilometres. This phenomenon is relatively rare in Australia, but has been known to occur in the lee of mountain ranges in parts of Tasmania, southern New South Wales and coastal north Queensland.

Land and sea breezes

Land and sea breezes are winds created by the daily heating and cooling of the land surface adjacent to the ocean. During the day, heating by the sun results in the land becoming warmer than the surface of the sea. The air over the land also heats up, becomes more buoyant and rises, reducing air pressure, and denser moist air over the ocean flows inland. This is the sea breeze. It generally starts in late morning and strengthens in the afternoon, and may reach a speed of about 35 km/h at the coast. Sea breezes rarely penetrate more than 30 km inland in Australia, as the coastal range blocks their progress. Occasionally, however, they may penetrate much further. For example, under suitable conditions sea breezes blow into Canberra, more than 100 km from the coast. Sea breezes can strengthen the prevailing wind, reduce its speed or even reverse it, depending on the direction and force of the two airstreams.

At night, the land cools and there is sometimes a flow that is the reverse of the sea breeze. This is the land breeze. It is generally less turbulent than the sea breeze and is frequently dry and warm. Its effect is less than that of the sea breeze, and is mainly concentrated on the coastal strip.

As they bring maritime air ashore, sea breezes are normally moist and the moisture content of dead fuel increases quite rapidly in response to the increased humidity. Fires in coastal regions are often subdued due to the resulting relatively high fuel moisture contents, and will burn in a highly predictable direction while the sea breeze persists. When sugarcane burning was normal practice in north Queensland, farmers would always wait for the onset of the sea breeze before burning-off.

Sea breezes may cause difficulties for firefighters in two situations. When there is a large pressure gradient between the land and the sea, a sea breeze may reverse the direction of the prevailing wind. On these occasions the sea breeze may arrive suddenly and unexpectedly with the characteristics of a frontal squall, causing a rapid change in, or even reversing, the direction of fire spread.

In parts of the country, such as east Gippsland in Victoria and south-western Western Australia, the dry continental wind blows directly off-shore. The resulting air mass over the ocean may have insufficient time to absorb moisture from the ocean before the sea-breeze circulation commences and blows it, still dry, back over the land. The rise in relative humidity and drop in temperature normally associated with a sea breeze does not occur, and firefighters do not get the relief they might expect.

Interaction of fuel moisture with relative humidity

The moisture content of dead fuel is determined by the relative humidity of the air, so that at night, when the humidity is around 100%, it is saturated with water. Dead fuel can absorb around 35% of its weight in moisture; this value is the saturation moisture content. As humidity falls during the day the moisture content of the fuel also falls, albeit more slowly, and as humidity rises in the evening the fuel's moisture content also increases, again more slowly than the humidity. If the relative humidity of the air were to remain constant, the moisture content of all dead fuel would eventually reach a corresponding steady value. This is the equilibrium moisture content.

Changes in the moisture content of the grass *Sorghum intrans* with relative humidity during the day are shown in Figure 6.12, and compared with the equilibrium moisture content for grass, over a range of humidities, determined in the laboratory. As the humidity drops in the early morning, the moisture content of the grass remains high until the free moisture from overnight dew is evaporated. The grass's moisture content is higher after 9 a.m. than the equilibrium moisture content because of the delay in moisture loss while the fuels are drying. As the humidity rises in the afternoon the grass takes up moisture and rapidly reaches the equilibrium moisture content. After 8.30 p.m. dew starts to form and the moisture content rapidly rises above the fibre saturation level.

The delay between a change in the relative humidity and a corresponding change in the dead fuel moisture content is called a hysteresis or, more commonly, a time-lag. The size of the fuel moisture time-lag depends on the coarseness of the grass

Local variation and erratic fire behaviour | 79

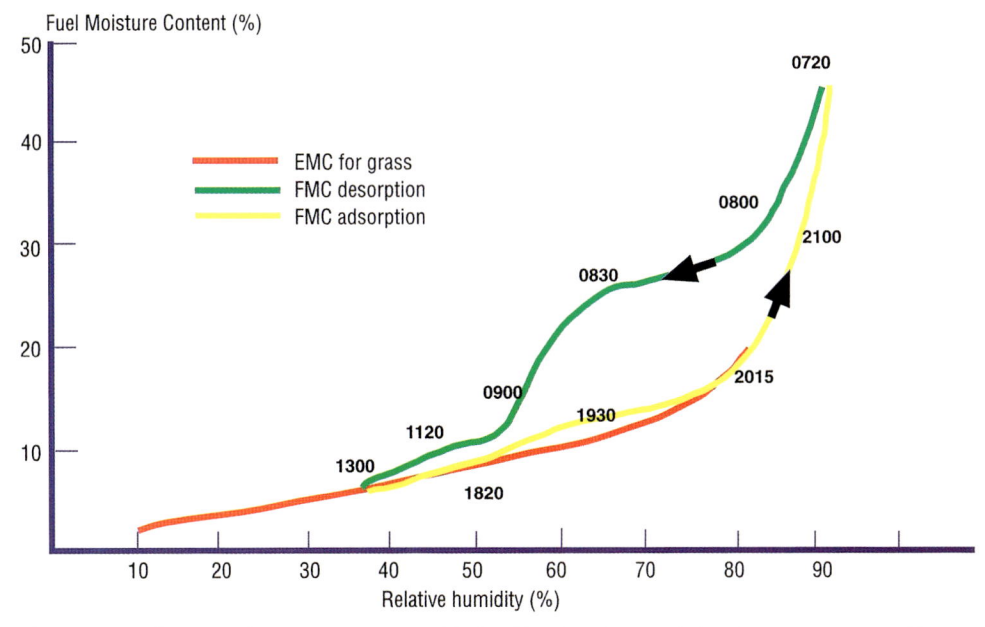

Figure 6.12: Change in fuel moisture content (FMC) of Sorghum intrans during the day compared with laboratory-determined equilibrium moisture content (EMC) for grass in relation to relative humidity.

stems and leaves and the rate at which relative humidity changes. Fine grass reacts much more quickly than coarse grass.

Generalised relationships between dead fuel moisture content, temperature and relative humidity have been developed (Fig. 6.13). These are useful for predicting the dead fuel moisture content of grass fuel, but assume normal diurnal trends in

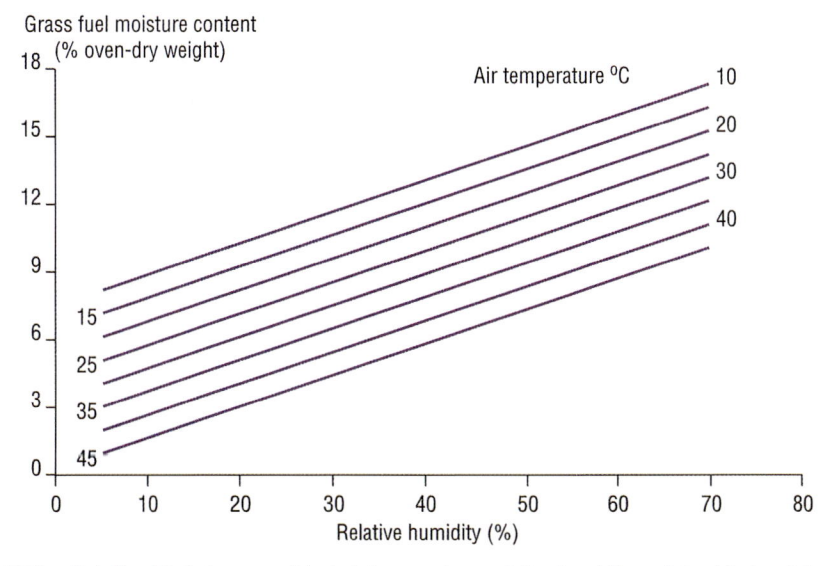

Figure 6.13: Relationship between ambient air temperature, relative humidity and dead fuel moisture content under field conditions.

temperature and relative humidity. The inclusion of temperature in the relationship takes into account some of the hysteresis effect on fuel moisture as the relative humidity changes during the day.

Any occasion that differs dramatically from the normal diurnal pattern of temperature and relative humidity on which the generalised relationships of Figure 6.13 are based will produce a moisture content different from that predicted, and hence different fire behaviour. For example, the presence overnight of a dry air mass with low relative humidities will produce fuel moisture contents the following morning that are lower than those predicted by Figure 6.13.

Litter fuel absorbs moisture much more slowly than grass fuel, and events of Ash Wednesday in 1983 illustrate the contrasting consequences. The cool change that day (see Fig. 6.3) brought a rapid rise in relative humidity and fires in grassland spread relatively slowly after its arrival – despite the extremely high winds associated with the change – due to rapidly increasing fuel moisture contents. In places, fires self-extinguished within a couple of hours of the change. On the other hand, litter fuels in forest absorbed moisture only slowly and forest fires spread extremely rapidly and continued to burn overnight.

This difference in moisture content of grassy fuel and forest litter can cause problems for firefighters in wildfire suppression. Where control lines have been established in grasslands near the boundary of forest blocks and it is necessary to burn-out the fuel between the wildfire and the control line, firefighters may find they cannot ignite the grassy fuel if the burning-out operation is delayed too long into the evening. If this happens, the firefighters will not be able to burn out all the fuel within the control line. As a result, the fire is likely to break away the next day if dangerous fire weather recurs.

Other factors that influence fire behaviour

Atmospheric stability

Atmospheric stability or instability can have a significant influence on fire behaviour. Most major fires burn under conditions of atmospheric instability, which allow the formation of a strong active convection column over the fire. This can lead to increased wind speed near the ground due to in-draughts, long-distance spotting and the formation of fire whirlwinds.

A stable atmosphere with a strong inversion layer above the fire can prevent the formation of a strong convection column, and all smoke will be trapped beneath the inversion layer. This is the condition generally sought for prescribed burning operations. Signs of a stable atmosphere are a hazy sky and relatively constant wind direction; the top of the inversion layer is quite obvious from an aircraft.

Indicators of an unstable atmosphere are obvious well before a fire breaks out. The sky is very clear and visibility is very good, particularly after a cool change, and whirls of dust start to rise early in the day. Whirlwinds or dust-devils are common, particularly in the arid zone; they are initiated by irregular heating of the ground. Cumulus clouds may be present if there is sufficient moisture in the atmosphere.

Fire whirls are narrow twisting plumes of flame. With small fires, they persist for only a few seconds before dying out and reforming. However, large fast-spreading fires can produce whirls that may reach 20 m or more in height and travel down the flank of the fire for several minutes. Fire whirls can pick up loose burning debris, such as clumps of cut grass or manure, and deposit them outside the fire perimeter, causing considerable trouble for suppression forces. More persistent fire whirlwinds commonly occur on the lee slopes of hills; the mechanical turbulence of the wind flow over a hill will frequently initiate them just beyond the crest. Although a fire may slow on the lee slope of a hill, spotting caused by whirlwinds can hamper fire suppression in the area.

The massive whirlwinds that can develop in forest fires or land-clearing burns are maintained by heavy fuels. These are not common in grassfires but have occurred where large fires have moved out of the forest onto grassland.

An extreme example of convective activity under unstable conditions is a thunderstorm down-draught. When a large cumulus cloud develops into a thunderstorm, rain falling within and below the cloud drags air with it, starting a down-draught. The rain also cools the air below the cloud base, and as it is colder and denser than the surrounding air it sinks very rapidly. When this descending air hits the ground below the base of the cloud, it blows radially outward in all directions (Fig. 6.14).

The surface winds from a thunderstorm down-draught start abruptly, can be very strong (up to 65 km/h), and usually last for around 15–30 minutes. Down-draught winds can easily reverse the direction of the prevailing surface wind more than 10 km from the storm centre. Firefighters should be very wary of wind changes when they are burning-off or fighting wildfires when thunderclouds are in the vicinity. An early warning sign that a down-draught wind may occur is when virga (a veil of rain beneath the cloud, that does not reach the ground) is observed below a large high-level thunderhead. Firefighters should expect a sudden wind change with wind blowing from the direction of the thunderhead, even when it appears to be far away.

Solar radiation

The main effect of solar radiation is to assist the fuel drying process. This can be in the form of curing dying plants or evaporating surface moisture deposited by high relative humidity. The easiest measure of solar radiation is the relative cloudiness

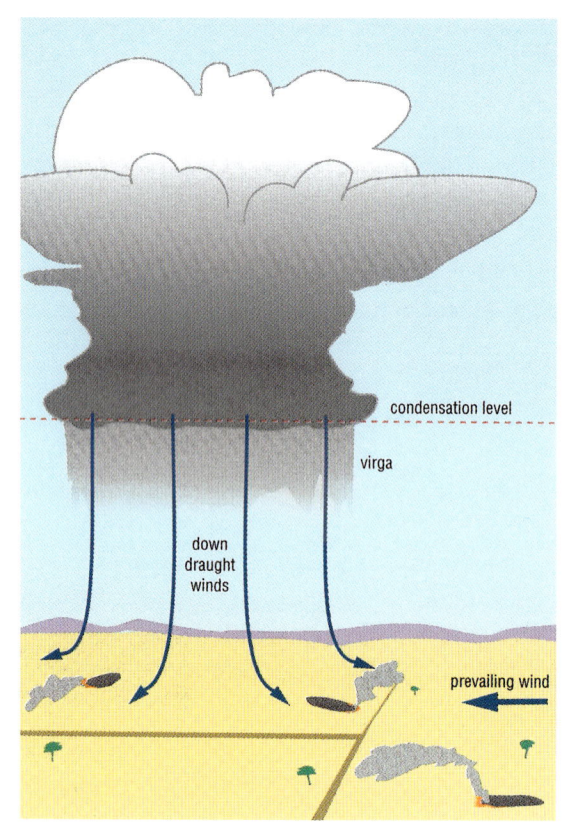

Figure 6.14: Thunderstorm down-draught winds are initiated by rapidly descending air below a thundercloud. At ground level they blow radially outwards and may change the direction of the surface winds up to many kilometres away.

(measured in octals, or eighths) of the sky. A reading of 8/8 is totally overcast, so solar radiation will be low, and 0/8 is totally clear, with high solar radiation.

Solar radiation will be greatest on very clear days with high atmospheric instability. The strong radiation will heat the ground which, in turn, further increases the instability of air near the ground and lowers fuel moisture contents. These two factors contribute to erratic fire behaviour.

7
FIRE DANGER

In Australia, the term 'fire danger' refers to a combination of weather and fuel conditions that indicate how difficult a fire will be to suppress. Because forest and grassland fuels have different burning characteristics, e.g. forest fuels will burn when grasslands are green and cannot burn, separate fire danger rating systems have been developed for forests and grasslands. Different grasslands have different fuel characteristics and, where these characteristics are extreme, such as between spinifex grasslands and arable pastures, different danger rating systems could be developed.

As discussed in Chapter 4, variation in fuel characteristics among different grass species makes little difference to the speed of fires in continuous pasture, although they may affect flame height and flame depth in ways that depend on the structure and compaction of the fuel. The CSIRO grassland fire danger rating system employs only one fuel variable – degree of curing. Combined with temperature, relative humidity and wind speed, this gives an index of the degree of difficulty of suppressing fire in a standard average pasture. The five fire danger rating classes – Low, Moderate, High, Very High and Extreme – represent the degree of suppression difficulty in such a pasture. These classes were defined by A.G. McArthur in 1966 and remain widely accepted.

Table 7.1 outlines the levels of suppression difficulty associated with different values of the fire danger index. At an index value of 1 or 2 fires will not burn, or will burn so slowly that control presents little difficulty. At an index value of 100 or more, fires will burn so hot and vigorously that control is virtually impossible.

Table 7.1: Fire danger index and difficulty of suppression

Fire danger index	Fire danger rating	Difficulty of suppression
0–2.5	Low	Low. Head fire stopped by roads and tracks.
2.5–7.5	Moderate	Moderate. Head fire easily attacked with water.
7.5–20	High	High. Head fire attack generally successful with water.
20–50	Very High	Very high. Head fire attack may succeed in favourable circumstances. Back-burning close to the head may be necessary.
50–200	Extreme	Extreme. Direct attack will generally fail. Back-burns from a good secure line will be difficult to hold because of blown embers. Flanks must be held at all costs.

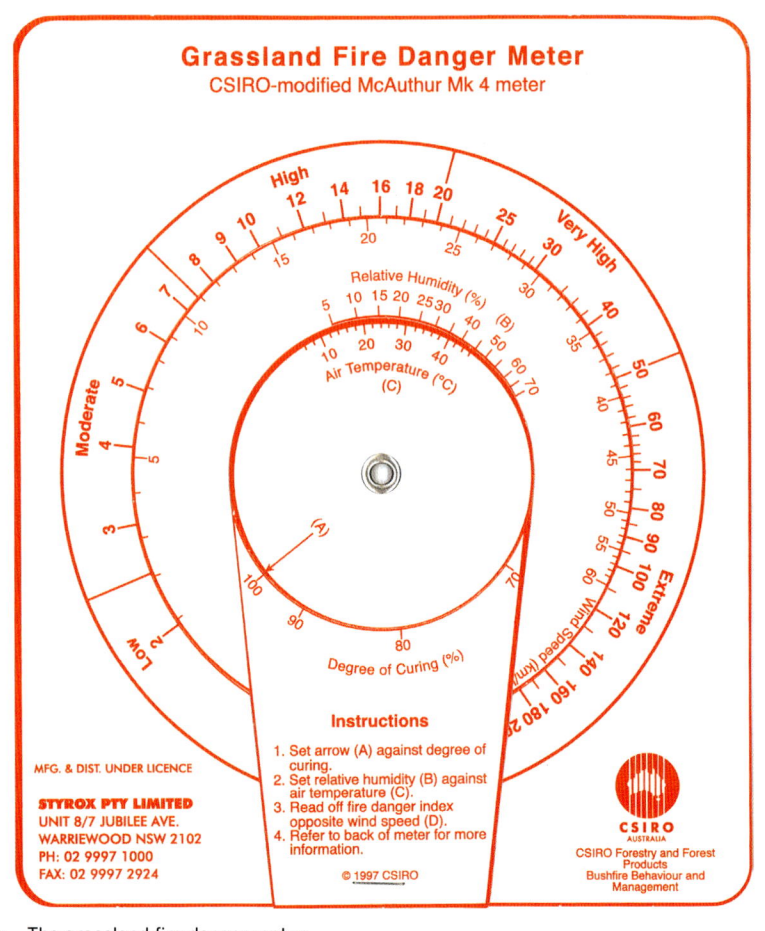

Figure 7.1: The grassland fire danger meter.

Fire spread and fire danger prediction

The conditions that affect relative fire danger do not always affect the rate of spread in the same way. Fire danger and difficulty of suppression are related exponentially to wind speed, i.e. as wind speed increases, the difficulty of putting out a fire rises at an ever-increasing rate. The rate of forward spread, on the other hand, has a near-linear relationship to wind speed. Thus, while wind speed is an important factor in predicting both fire spread and fire danger, fire spread cannot be directly linked to a fire danger index.

The CSIRO fire danger rating system uses the same relationships as the McArthur Mk 4 fire danger rating system, with two important differences. First, rate of spread has been removed. It should be calculated separately using the CSIRO fire spread prediction system. Second, the index value can exceed 100 – the index is open-ended. In McArthur's original system, an index value of 100 was intended to represent the 'worst possible' fire weather conditions likely to be expected in Australia. This value has been exceeded on several occasions since 1966.

McArthur's system has been used by rural fire authorities across Australia for more than 40 years, and his fire danger classes have been found satisfactory for providing public warnings, setting preparedness levels and generally providing a good indication of the difficulty of fire suppression over a wide range of conditions. Retaining the same fire danger rating system will make it possible to compare contemporary fire weather with historical records. Separating fire danger from fire spread will allow us to make adjustments to the fire spread prediction system in the future without disrupting organisational arrangements based on fire danger levels.

The amount of fuel present obviously affects fire suppression difficulty – if there is no fuel there is no fire danger. However, it is difficult to include fuel load in a fire danger rating system designed for application at a regional level. Thus, for general forecasts it is necessary to assume a standard fuel condition. In exceptional circumstances – where fuels are absent or heavily eaten-out across the whole region – sensible adjustments can be made by local fire authorities in setting preparedness levels and providing public warnings on the level of fire danger.

Similarly, difficulty of fire suppression depends on the resources available. For example, one person may find it extremely difficult to suppress a fire under conditions of moderate fire danger, even in sparse fuels. In regions where suppression resources are limited, fire authorities may need to provide public warnings and declare total bans on the lighting of fires at lower values of the fire danger index than are used in more closely settled regions.

8
WILDFIRES AND THEIR SUPPRESSION

In the pastoral areas of south-eastern and south-western Australia almost every summer is likely to bring several days of extreme fire weather and the accidental ignition of fires. If fires are not controlled soon after ignition in conditions of strong dry wind, high temperature and abundant continuous fuel, they will travel at high speed for several hours, burning out huge areas. Control of the head fire is impossible, and the area that is burnt largely depends on the time a fire starts, the period that elapses before a change of wind direction and the weather conditions after the change.

The damage wrought by these huge grassfires can run into millions of dollars. The biggest costs are incurred in replacing fencing and losses of stock and feed. Many urban people may not appreciate that all the dead dry grass in paddocks during summer is dry feed for stock and represents graziers' livelihood. Even when stock losses are low, a grazier may be ruined financially by a large fire simply because of inability to feed stock or agist them elsewhere for the remainder of the summer. For this reason, graziers prefer to suppress fires directly in order to save what grass they can on their properties, rather than burn-out from roads or firebreaks.

The synoptic situations that produce extreme fire weather are well-known. In south-eastern Australia, extreme days occur when a strong blocking high-pressure system is located in the Tasman Sea ahead of a strong cold front moving across the Great Australian Bight (Fig. 8.1).

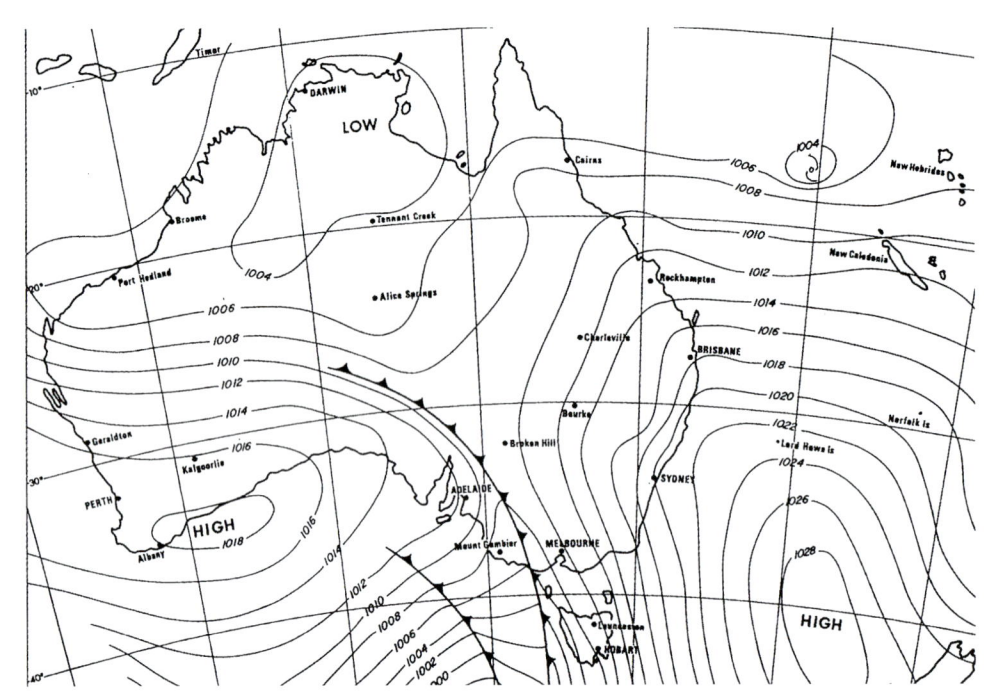

Figure 8.1: Synoptic situation at 3 p.m. EDST, 12 February 1977. Extensive grassfires occurred in western Victoria. Source: McArthur *et al.* (1982).

Atmospheric circulation around the high-pressure system brings hot dry air from central Australia down through South Australia, southern New South Wales and Victoria as a strong northerly wind ahead of the cool change. As the change passes, there is a sudden shift in wind direction to the west or south-west behind the front (see Fig. 6.3). Although the winds behind the change are cooler and moister than the preceding northerly winds, if there is no rain associated with the change grassfires will continue to burn strongly for several hours until the fuel takes up moisture. Huge areas can be burnt after the change if a fire has travelled a long distance under the northerly wind and there has been no effective suppression on the eastern flank. Elsewhere in Australia, extreme fire days will occur under a synoptic situation that causes strong winds to blow dry air from the centre of the continent.

In this chapter we give examples of large past grassfires, and discuss options for suppression under very high to extreme fire danger conditions.

Past wildfire events

Mangoplah, New South Wales, 22 January 1952

This fire (Fig. 8.2) burnt from Mangoplah, north of Holbrook, NSW, to Corryong in north-eastern Victoria, a distance of 98 km, during two consecutive days of extreme fire danger. The area burnt, mostly on 25 January, was more than 330 000 ha.

Wildfires and their suppression | 89

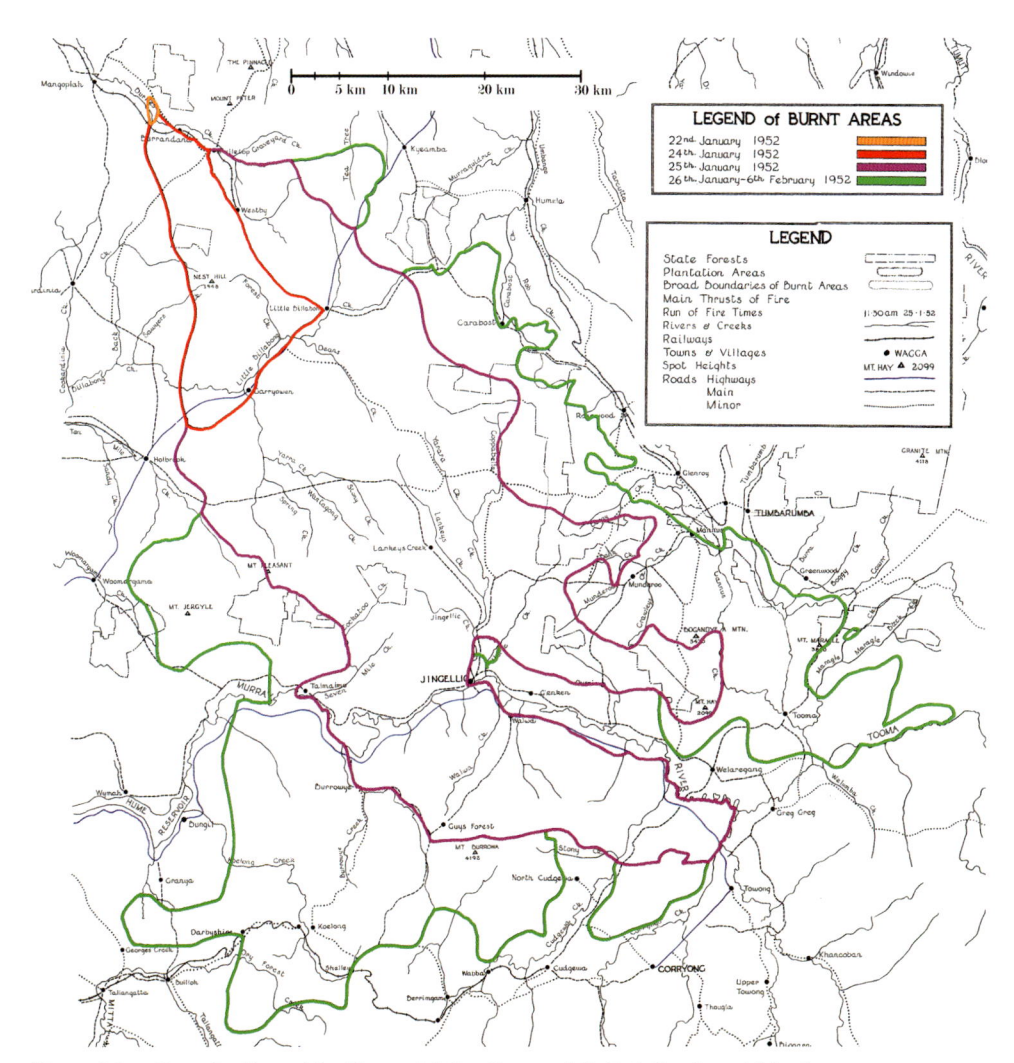

Figure 8.2: Spread pattern of the Mangoplah fire. Source: A.G. McArthur (unpublished).

The fire commenced on 22 January from fettlers burning-off on The Rock–Westby railway line near Mangoplah. This fire was brought under control by local bushfire brigades after burning an area of 150 ha. However, on 24 January, under extreme conditions (temperature 42.5°C, relative humidity 29%, wind speed 40 km/h, grassland fire danger index (GFDI) 60), the fire broke away due to sparks from a stump reported to be 350 m inside the burnt country. It spread under a strong north-westerly wind and by midnight had burnt an area estimated at 27 000 ha. The wind, although abating, continued to blow throughout the night and, combined with very hot dry overnight conditions (the overnight minimum at Wagga Wagga was 29.5°C), thwarted attempts to hold the fire on the Hume Highway near Garryowen.

On 25 January the mean wind speed increased to around 48 km/h, still from the north-west. With a temperature of 41°C and a relative humidity of 15%, the maximum GFDI value reached 115. The fire crossed the Murray River near Jingellic at 10.30 a.m. and burnt in a south-easterly direction for another 13 km before a cold front with a westerly wind change passed through the area at 11.30 a.m. By midnight the fire had burnt much of the total area eventually affected, and most perimeters in grasslands had been controlled. However, the fire continued to burn in timbered country until 10 February before it was brought under total control.

The Mangoplah fire illustrates the enormous areas that can be affected when a fire burns over two consecutive days of extreme fire danger. The wind did not abate sufficiently at night to allow firefighters to bring the fire under control on the Hume Highway and, although the organisation of firefighters that night was described as chaotic, they were always going to have difficulty controlling the fire along a tree-lined road while the wind continued to blow. Firefighting resources and equipment have improved since 1952, but a similar or greater area can still be burnt over two consecutive days of extreme fire weather.

Western District of Victoria, 12 February 1977

On this day some 70 fires started throughout Victoria. Although most were controlled to less than 1000 ha, 11 major fires burnt 103 000 ha. Only three of these burnt more than 10 000 ha each. Nine of the 11 major fires were started by sparks from power lines. The largest, the Wallinduc–Cressy fire (Fig. 8.3), burnt 39 200 ha and resulted in the death of three people and the destruction of 39 houses and other buildings.

Conditions during this fire were such that grasses were fully cured and had been for some days. The fire started at 1.32 p.m. and for much of the afternoon a north-north-westerly wind blew up to a maximum speed of 50–55 km/h, producing a maximum GFDI of 98 (temperature 36°C, relative humidity 22%). Sustained rates of spread in the order of 17 km/h were recorded for extended periods (up to 30 minutes), with a mean spread rate of 14 km/h over two and a half hours. Three hours after ignition the fire had travelled about 34 km and burnt some 16 700 ha. Then a south-westerly cold front passed through the area, with wind speeds of around 40 km/h. The entire eastern flank took off, initially at 13 km/h, and over the next two and a half hours burnt a further 22 500 ha before the fire became controllable and was extinguished.

Another of the day's fires, the Tatyoon–Streatham fire (Fig. 8.4), started 14 minutes before the Wallinduc–Cressy fire and 70 km to the north-west. It sustained a maximum rate of spread of 19 km/h over 30 minutes, but had a slightly lower mean rate of spread (12 km/h) than the Wallinduc–Cressy fire and burnt a total of 20 100 ha. One man was killed, 38 houses were destroyed and the small township of Streatham was severely damaged when 22 structures were burnt.

These two fires show the importance of effective suppression on the eastern flank. Both fires had burnt about the same area (15 000–16 000 ha) prior to the wind change. On the Tatyoon–Streatham fire, firefighters were able to control

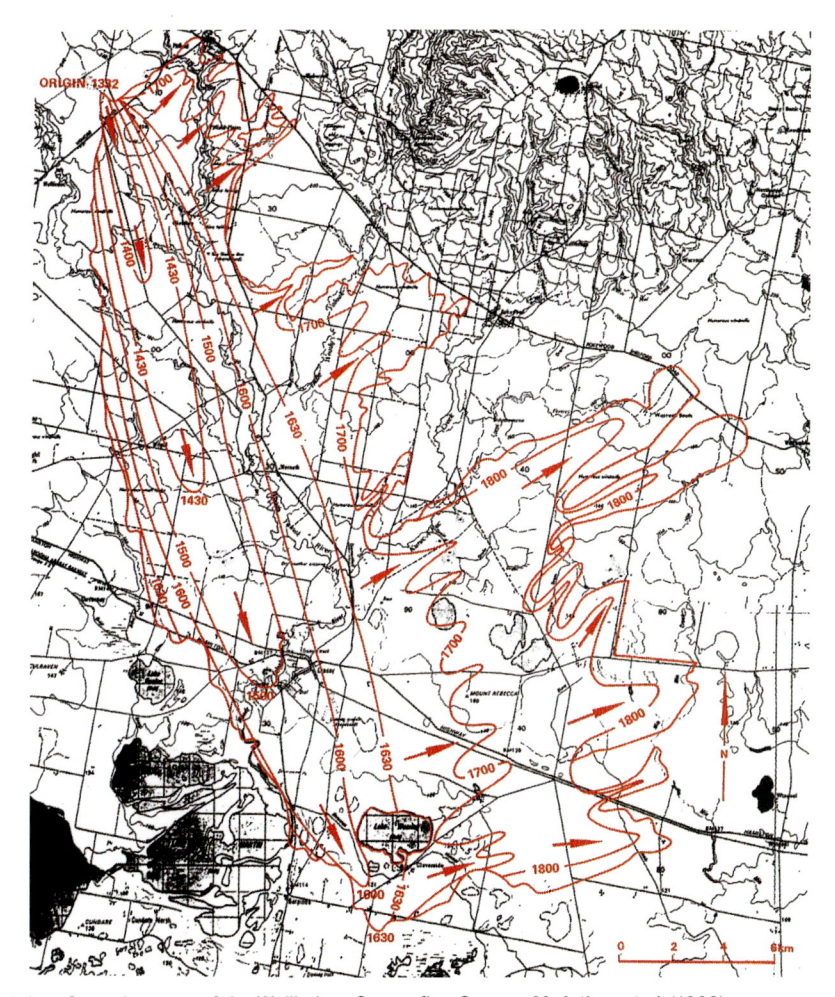

Figure 8.3: Spread pattern of the Wallinduc–Cressy fire. Source: McArthur *et al.* (1982).

22.5 km along the eastern flank before the wind change and thus limit the area burnt after the change to only 5000 ha. In contrast, fire suppression on the eastern flank of the Wallinduc–Cressy fire before the wind change was hindered by rough stony country near the origin and by the fact that local firefighters did not receive support from additional brigades from the north and west (who were already engaged on fires to the west which had started earlier). Firefighters were able to control only 1.5 km of the eastern flank, and as a result more than 22 500 ha burnt after the wind change.

South Australia and Victoria, Ash Wednesday, 16 February 1983

Ash Wednesday, 16 February 1983, brought the worst fire disaster in Australia since Black Friday, 13 January 1939. About 370 000 ha of forests and grasslands was burnt, 76 people killed and some 2500 structures destroyed. Ash Wednesday is a prime

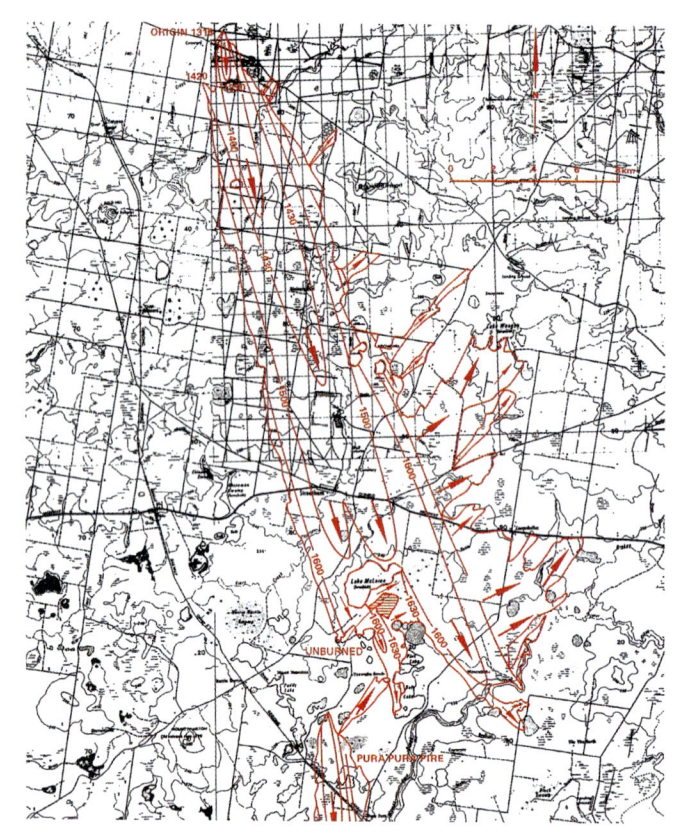

Figure 8.4: Spread pattern of the Tatyoon–Streatham fire. Efficient suppression held 22.5 km of the eastern flank before the wind change and dramatically restricted the total area burnt.
Source: McArthur *et al.* (1982).

example of particularly severe fire weather conditions in south-eastern Australia. Timing of the passage of the cold front was such that, as it swept across southern Australia during daylight hours, extreme fire weather extended from Port Lincoln in South Australia to east of Melbourne, Victoria – a distance of 800 km. At most locations, hot strong northerly winds started blowing early in the morning (9 a.m. DST) and increased to average mean speeds of 45–50 km/h for several hours preceding the front. Unusually strong westerly winds were associated with the frontal change, which reached Ceduna at 12.30 p.m., Adelaide at 2.45 p.m. and Melbourne at 8.30 p.m. Mean wind speeds exceeded 70 km/h, with gusts up to 110 km/h.

Much of south-eastern Australia was experiencing severe drought at the time (see Fig. 2.9), and most of the damage occurred in forested areas. However, a number of major grassfires also occurred in South Australia and Victoria in areas where the drought was not extreme. The pastoral area of south-eastern South Australia was one area with abundant grassy fuel. Two major fires (the Clay Wells and Narraweena fires) started in these grasslands and burnt parallel to each other

Figure 8.5: Origin and burnt area of the Tatyoon–Streatham fire. Photo: CFA, Victoria.

through grassland and pine plantation (Fig. 8.6). The Narraweena fire was the larger of these, and burnt for four hours with an average rate of spread of 18 km/h. After the wind change, the two fires swept together very rapidly and the southern end of the Narraweena fire entered conifer plantations more than 65 km from its origin.

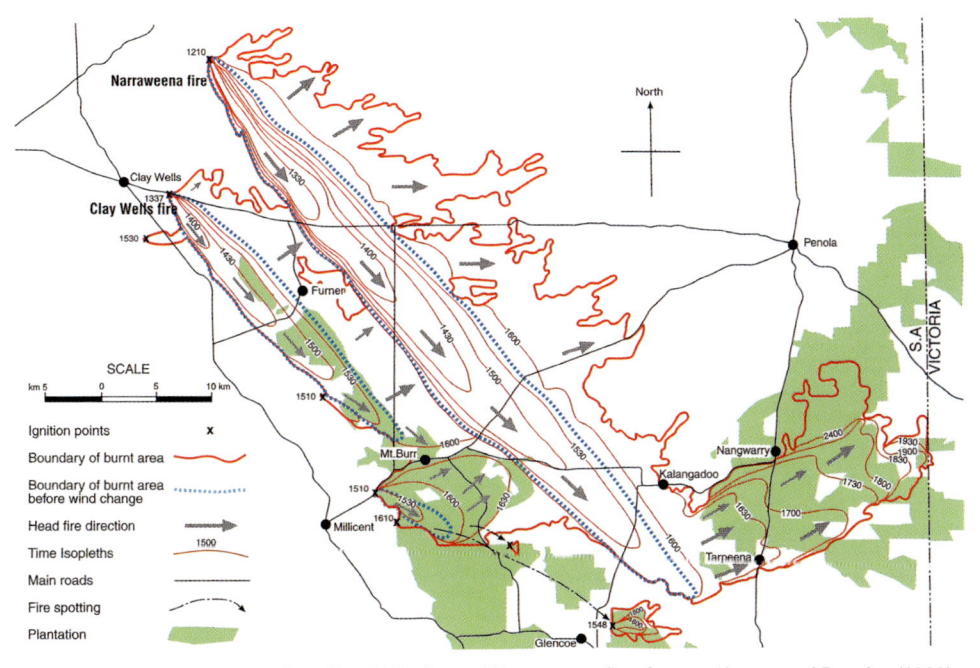

Figure 8.6: Spread pattern of the Clay Wells fire and Narraweena fire. Source: Keeves and Douglas (1983).

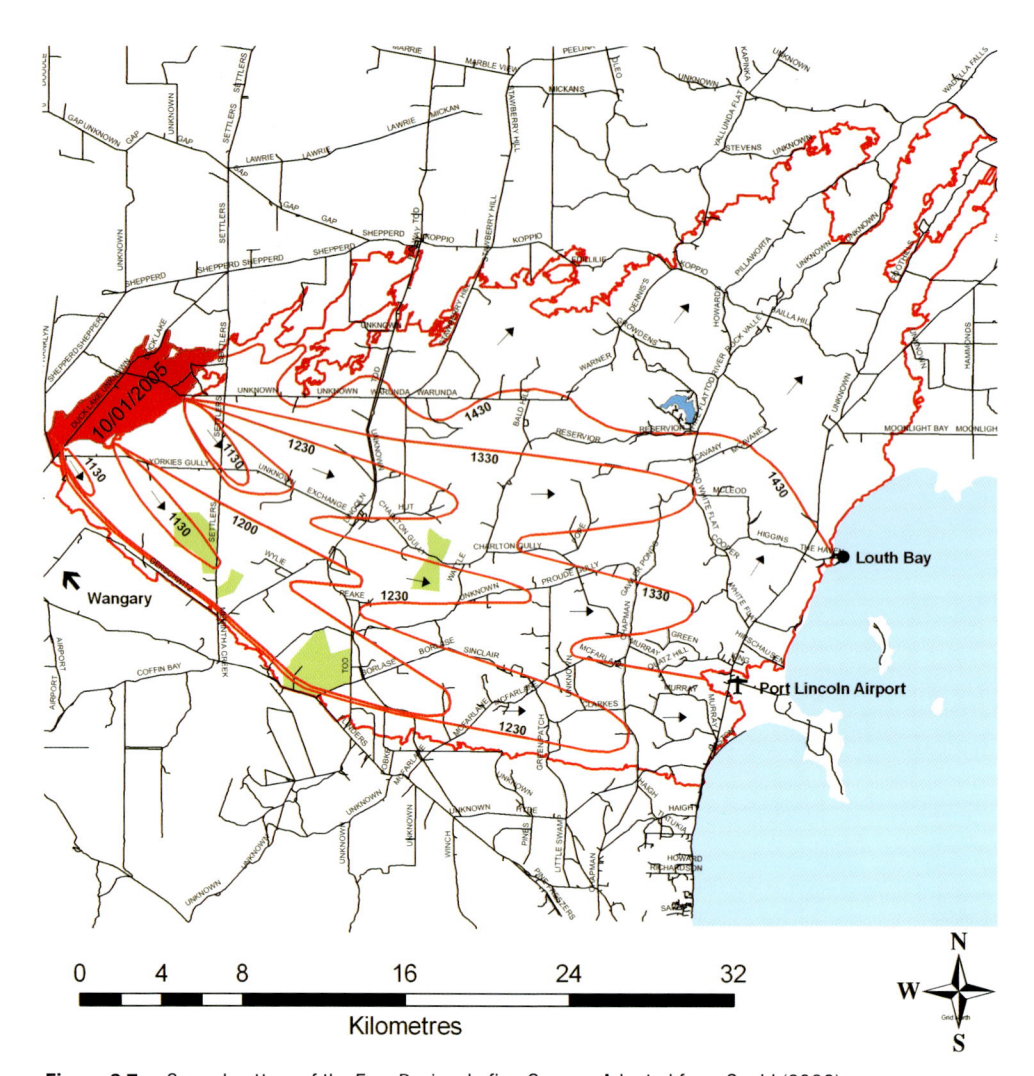

Figure 8.7: Spread pattern of the Eyre Peninsula fire. Source: Adapted from Gould (2006).

As winds abated and humidity rose two hours after the arrival of the front, the fires in grassland were brought under control. However, those in forests continued to burn throughout the night and were not completely controlled for another two days.

The extremely strong winds associated with the frontal change extinguished, by blowing out, large sections of the western flanks of both the Clay Wells and Narraweena fires. The eastern flank of the Narraweena fire did not travel as far as might have been expected considering the strength of the wind immediately behind the change. Most of the evidence suggests that this reduced rate of spread was not due to the very high wind speeds but rather to increasing fuel moisture content caused by the rapidly increasing humidity and, in places, scattered light rain behind the front.

Other major grassfires on Ash Wednesday were at Clare and in the Adelaide Hills, South Australia, and in the Western District of Victoria.

Eyre Peninsula, South Australia, 10 January 2005

The Wangary fire of 10 and 11 January 2005 (Fig. 8.7) burnt around 78 000 ha of the lower Eyre Peninsula, claimed nine lives (including those of four children and two firefighters) and destroyed 93 houses. The fire is important not only because of the extreme fire behaviour and the extensive damage but also because the lower Eyre Peninsula is an area where large fires occur infrequently because of the moderating effect of the sea on the western side.

The fire started from an accidental ignition on a roadside at around 3.15 p.m. on 10 January near Wangary on the western side of the Eyre Peninsula. Although the weather conditions at Port Lincoln were extreme, the weather in the fire area was milder and the fire danger high to very high. The fire was checked with an area of around 1500 ha but about 7 km of the eastern perimeter had burnt into a dry swampy region containing areas of bare salty ground separating large dense tussocks of sedge. The flames self-extinguished over a ragged edge but fuel in the swamp continued to smoulder overnight. The fire was not surrounded by a bare earth break and only limited mop-up was carried out.

By 9 a.m. on 11 January the fire danger was extreme and fire soon broke away at several locations along the eastern flank. At Port Lincoln the fire danger was extreme for eight hours and peaked at 12 p.m. when 10-minute weather associated with the passage of a frontal change was temperature 42°C, relative humidity 3% and mean wind speed 61 km/h. This produced a peak fire danger index of 369, the highest grassland fire danger index ever recorded. However, the average rate of spread was 17 km/h although there were reported short periods of spread estimated at over 30 km/h for 10 minutes associated with extreme wind speed accompanying the frontal change. Fatalities occurred at two sites at approximately the time of the wind change. They were associated with this period of extreme fire behaviour, when the rate of spread appeared to double for around 10 minutes.

The Wangary fire illustrates the difficulty of maintaining fire awareness in areas where serious fires are infrequent. Post-fire interviews of people in the area indicated that although they were aware of fire safety advice they were not confident of putting it into practice. In addition, the exact timing of a frontal change is impossible to predict in the field and the violent fire behaviour that is usually associated with the change makes survival in the open very very difficult. In this area of South Australia vestiges of the original vegetation remain mainly along the roadside verges, increasing threat to travellers overtaken by fire.

Suppression of grassfires under extreme conditions

Initial attack

During extreme fire weather, a fast and concentrated attack soon after a wildfire starts provides the best chance of bringing it under control. Firefighters should aim to reach the fire while it is still narrow and before it reaches its full potential rate of spread. Several tankers must arrive within 15 minutes of the fire starting. Under these circumstances there may be opportunities to bring the head under control if it is briefly checked by firebreaks, areas of sparse fuel or ploughed/fallow paddocks.

The firefighters arriving first at the scene should take a few minutes to appraise the situation before launching into direct suppression of the fire. If the whole fire can be seen at once, it may be possible to attack the head fire. Firefighters should check to see if the head fire has been held up for some reason and estimate the resources that will be needed to bring the fire under control. They should be careful not to be overconfident and thus underestimate the resources needed, even if the fire looks relatively mild. A mild appearance may be due to a temporary lull in the wind or an area of sparse fuel, either of which may provide a brief opportunity to attack the fire directly. Additional firefighters will be needed if the situation changes, and will in any case ensure a speedy mop-up if the fire is confined to a small area.

If the whole fire cannot be seen at once, it is unlikely that firefighters will be able to suppress the head fire. It is highly probable that they will be faced with an extended suppression effort, and considerable resources will be required. A systematic attack should commence near the origin, to provide an anchor point to which firefighters can safely retreat should the wind direction change. Fire suppression should then progress along both flanks in order to restrict the width of the fire. If the head fire is checked for any reason, the suppression forces may be able to catch up to it and bring it under control.

Extended attack

Once a fire has reached its potential rate of spread, it is highly unlikely that the head fire will be stopped by any suppression tactics until it runs into a very substantial barrier. It is now most important to concentrate on the fire flank that is most likely to break away when the wind direction changes. In eastern Australia, this is the eastern or north-eastern flank. The normal pattern of wind shift during the day is anti-clockwise, i.e. the wind will back from the north to the west. This will make the eastern flank more difficult to attack than the western flank, but suppression efforts must concentrate on this flank even if firefighters can take effective action only during lulls or when the wind veers towards the north. In many situations, the area burnt by the fire will not alter much whether or not any suppression effort is undertaken on the western flank before the arrival of a frontal wind change. Nevertheless, some systematic suppression should commence on the western flank. Graziers with properties on that side will want to minimise the amount of grass burnt on their properties and it is important that they join an organised suppression effort.

In some areas, the expected wind change may come from a sea breeze rather than a frontal change. In the Canberra region, the sea breeze has occasionally been strong enough to overcome the prevailing westerly wind and suddenly reverse the direction of fire spread. Fire controllers can usually obtain good forecasts of the onset of a sea breeze from the Bureau of Meteorology, particularly if they provide Bureau forecasters with information about temperature and prevailing wind at the fire site.

Head fire attack

If an attack on the head fire is successful, a fire can be brought under control rapidly and the area burnt kept to a minimum. However, head fire attack is difficult and dangerous under conditions of very high or extreme fire danger. Experienced crews with a good knowledge of fire behaviour are required to undertake it safely and successfully.

When a fire is too intense to attack the head directly, the best chance of success is to hold it at a road or firebreak. Sufficient forces must be available to control spot fires when the fire reaches the break. In the case of relatively slow-moving fires, the width of the break can be extended by back-burning on its upwind side. The back-burn will spread only slowly into the wind and will not be drawn towards the main fire. The back-burn can be extended by lighting additional fires between it and the main fire, but this is an extremely hazardous operation as firefighters may become trapped between the two fires. When the head fire reaches the break and the convective centre moves across it to the downwind side, strong winds behind the flame front will cause the flames to lean over parallel to the ground and firefighters on the break will be subjected to high levels of smoke and convective heat (Fig. 8.8). These winds, from immediately behind the head fire, will blow sparks, ash and debris across the break for a few minutes. The break must be wide enough to allow firefighters to work safely outside their vehicles and suppress any spot fires across the break as soon as they occur.

Under extreme conditions, head fire suppression is rarely possible and back-burning to increase the width of a break is neither safe nor feasible. A fire travelling at 18 km/h is covering 300 m per minute. As it takes at least 20 minutes to organise crews and establish a safe wide break, firefighters would need to be in the correct position when the fire was 6 km away. Even if they could hold their back-burn under these conditions, there is a strong possibility that the main fire would miss the back-burn altogether (Fig. 8.9). The crews would become disoriented and disorganised and the back-burn would escape from control, increasing the overall width of the fire. The only safe and effective strategy under these conditions is to undertake a well-organised suppression effort on the flank of the fire to restrict the area burnt when the inevitable wind change occurs. The temptation to attack the head should be strongly resisted.

Figure 8.8: Experimental fire B111, NT, 6 August 1986. Flames are blown directly over a firebreak by convection-enhanced winds behind the fire.

Flank fire attack

As discussed above, a well-organised attack on the eastern flank is the best strategy for fast-moving grassfires burning under extreme conditions (Fig. 8.10). This attack must commence at an anchor point, i.e. a place at the rear of the fire that will be secure at all times. The attack must be well-organised, with plenty of resources available to come in behind the lead tankers, mop-up and hold the extinguished perimeter. The lead tankers should resist the temptation to jump ahead of the back-up units if, say, a portion of the flank is held up by a road or break running parallel to the flank. Many hours of work can be lost in a few minutes if a fire reignites and breaks away near the rear of the flank and forms a secondary head fire that burns parallel to the flank. There is also a risk that the breakaway may trap the lead tankers.

It is a good tactic to commence grading a bare earth trail as soon as possible along any perimeter extinguished by water. Firefighters should work outside the fire perimeter and turn the graded material away from the fire edge. This avoids creating a mixture of smouldering material, soil and unburnt grass, which can be extremely difficult to mop-up.

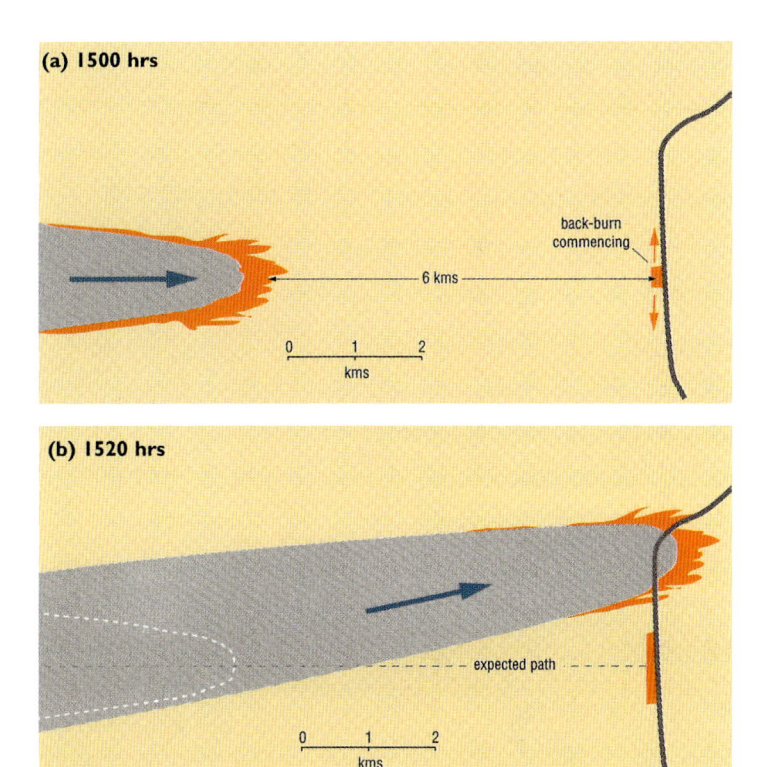

Figure 8.9: Back-burning against fast-spreading head fires is rarely successful. (a) A fire travelling at 18 km/h. Back-burning commences at 3 p.m., directly downwind of head. (b) At 3.20 p.m. a 1 km back-burn has been established but a 10° shift in the wind direction has blown the fire past the back-burn.

While flank attack is possible in open grasslands under extreme conditions, because of spotting flank attack is well-nigh impossible when a fire burns into patches of trees or wooded areas. Suppression along the flank will be held up until the patch is burnt out. Likewise, tree-lined roads are often ineffective as firebreaks in controlling either the head or the flank of a fire. If the opportunity arises, fire-fighters should always try to suppress a fire in open paddocks rather than wait for it to burn up to a tree-lined road.

Indirect attack and burning-out

Although direct suppression of the fire edge is the normal suppression method, indirect attack may be employed in certain circumstances. Firefighters may plan to use a road or firebreak parallel to the flank fire as the control line and burn-out the fuel between the break and the fire. Great care should be taken when burning-out with the wind and lighting lines of fire. The intensity of the head fire developing from the burning-out operation can be similar to that of the wildfire and may threaten firefighters downwind on other sections of the fire perimeter.

Figure 8.10: The head fire of the Narraweena fire, 16 February 1983, approaches Penola–Robe Rd, 90 minutes after ignition. The fire is already 2 km wide and back-burning is clearly impossible. Photo: D. Page.

Indirect attack will be necessary if a grassfire has entered a patch of bushland where direct attack is unsafe. Burning-out of bushland areas should be delayed until weather conditions abate. During conditions of extreme fire danger, burning-out (particularly with line fires) may cause long-distance spotting and considerably increase the overall suppression problem.

Mop-up and patrol

On most fires, mopping-up is the most important part of fire suppression. Some firefighters may enjoy the excitement of putting out the flames and be reluctant to undertake the long dirty task of mopping-up smouldering material close to the fire perimeter. It is essential, however, that mopping-up be done thoroughly and completely, particularly on small fires in the days leading up to days of extreme fire danger.

Mop-up requires a balancing act between letting the material inside the perimeter burn away and using water to put it out. If conditions are not dangerous, it can be a good tactic to allow as much material as possible to burn away before mopping-up. In the end, however, all the smouldering material must be mopped-up until the whole fire perimeter is dead-out to a distance of 100–200 m or more.

Logs or piles of smouldering grass, e.g. hay bales, can be pushed well into the fire area, scattered, broken up and extinguished with water. Hollow trees may need

Figure 8.11: Breakaway from a windbreak near the origin of the Pura-Pura fire, Victoria, 12 February 1977, after the wind change. This illustrates the importance of mop-up in the vicinity of trees. Photo: CFA, Victoria.

to be felled, cut up and split to get to the fire inside. The most reliable way to ensure that fuel is dead-out is to feel it with bare hands.

Special attention should be paid to any trees near the fire perimeter, and there should be regular sorties back along the flank to the origin to check for smouldering material (Fig. 8.11). It is essential that the fire boss designates sectors around the fire perimeter and assigns mop-up crews to specific sectors.

Under drought conditions, smouldering combustion can be astonishingly persistent and give little visible indication of its presence. Dead tree roots can burn underground for many days and fire may eventually break out on the surface many metres from the stump (Fig. 8.12). Large tussocks and peaty material can also smoulder for several days. The 1977 Penshurst fire in Victoria ignited from smouldering tussocks in an area of peaty soil that had been burnt two days previously. The strong dry wind of a day of extreme fire danger can rekindle the smallest ember. During the 1985 Cavan fire in southern New South Wales, smouldering dung in a sheep camp reignited and was blown more than 60 m across burnt ground. The fire broke away from control despite intensive efforts by fire crews.

Patrol does not end on the day of the fire. Crews should revisit burnt areas each day for up to a week if necessary to ensure that smouldering material has not rekindled. On days of extreme fire danger, crews should be stationed at any previous fire until they are required elsewhere. Often, only hot dry winds will cause the tell-tale wisps of smoke that indicate some fuel is still alight.

Figure 8.12: Burnt-out red box stump during the Cavan fire, NSW, 28 February–2 March 1985. One root burnt underground for 10 m and the fire resurfaced near the vehicle in the background.

In summary, mop-up involves breaking up concentrations of fuel, letting them burn-out as much as possible, dousing with water and feeling with bare hands. Mopping-up is definitely not the fun part of firefighting, but it is essential that it be done thoroughly and completely. As with most important jobs, the public will not know when it has been done well but they will know when something goes wrong.

Firebreaks

In parts of Australia, a lot of effort is put into constructing and maintaining firebreaks along roadsides and around property boundaries. These are usually designed for two purposes – to prevent burning-off operations or roadside fires from escaping, and to assist suppression of the head or flanks of a wildfire. Before putting considerable effort into creating a firebreak, landholders should consider the purpose for which it is being established and the conditions under which it will remain effective.

Firebreaks are very effective in grassy fuels provided the fire is not spotting. For example, as Figure 8.13 shows, a firebreak 10 m wide has a 99% chance of holding a 10 MW/m fire, i.e. a fire in 4 t/ha fuel travelling at 5 km/h. The same firebreak will hold faster-spreading fires in lighter fuels, but once the wind speed exceeds 25 km/h burning material will be blown along the ground and even quite wide breaks will be ineffective.

Firebreaks are much less effective if the grasses contain large seed heads that can spot ahead of the fire (e.g. *Phalaris* spp.) or there are trees nearby (Fig. 8.14).

A single tree close to a firebreak will decrease the effectiveness of the break under even relatively mild conditions (Fig. 8.15). Many breaks are located close to

Wildfires and their suppression

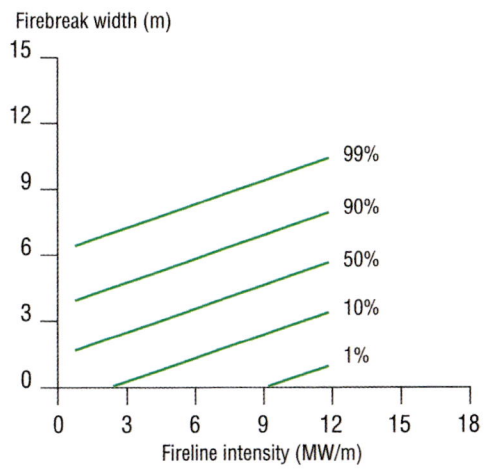

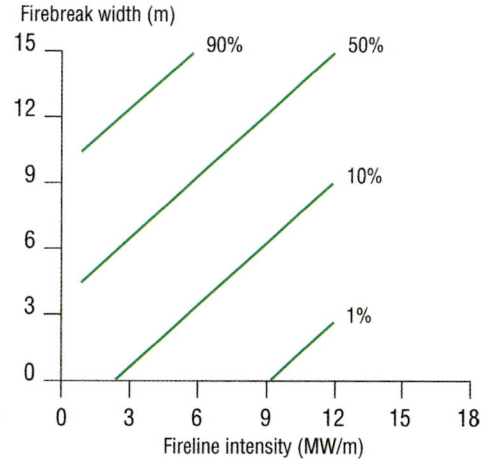

Figure 8.13: Probability of a firebreak holding a head fire in relation to fire intensity and firebreak width, where there are no trees within 20 m of the break. Source: After Wilson (1988).

Figure 8.14: Probability of a firebreak holding a head fire in relation to fire intensity and firebreak width, where there are trees within 20 m of the break. Source: After Wilson (1988).

tree-lined roads. These may be suitable for containing fires within the paddock, but will be largely ineffective in containing fires starting on the road verge. Where there is a serious problem with roadside fires starting in a tree-lined verge, either the verge should be burnt regularly prior to summer or a firebreak should be located inside the adjoining paddock some distance from the trees.

Figure 8.15: Origin of the Junee fire, NSW, 3 January 1990. This break confined the fire to the road verge until firebrands from the tree blew across it.

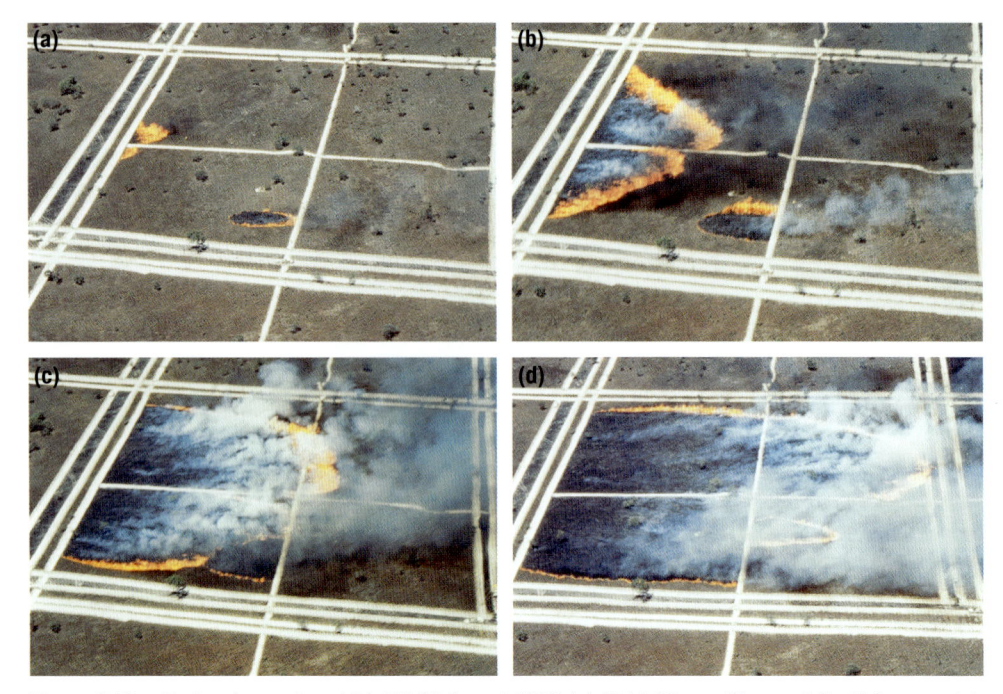

Figure 8.16: Firebreak experiment F4, NT, 20 August 1986. (a) 12.14:29 p.m. The small fire lit from a point is approaching the 1.5 m-wide break. Ignition of a second fire from a line 150 m long is commencing. (b) 12.15:05 p.m. The small point fire is held up early in its development by the 1.5 m break. The line fire represents the full potential for fire spread in the prevailing conditions. (c) 12.15:34 p.m. The point fire is still held by the 1.5 m break. The burnt-out area of the spot has blocked most of the lower head of the line fire. A single spot fire has carried across the break. The upper head has swept across the 1.5 m-wide break over the entire width of the head fire. (d) 12.16:09 p.m. The spots from the lower head are developing. The upper head is being held by a firebreak 10 m wide. The lateral firebreak through the centre of the block is continuing to hold the right flank of the upper head fire.

Under strong winds typical of days of extreme fire danger, even firebreaks more than 40 m wide may be quite ineffective in stopping a head fire. However, breaks more or less parallel to the prevailing wind direction may hold the flank fire and assist suppression both directly and by keeping the fire narrow, thus reducing its speed (Fig. 8.16).

Bare-earth firebreaks are essential around any burning-off operation, even where it may seem easy to control the fire directly with water. The firebreak will not only contain the fire but will reduce the effort required for mopping-up once the burn is completed. Bare-earth firebreaks are also essential around farm buildings and other flammable assets, even when the surrounding grasslands have been heavily eaten-out. Under extreme conditions, a fire with flames only a few centimetres high can spread across eaten-out paddocks. The firebreak will help prevent the fire burning into flammable materials near the homestead.

9

GRASSFIRE INVESTIGATION

At first glance, a burnt paddock is a burnt paddock with nothing to see but ash. However, an understanding of how grassfires burn can help in the recognition of fire patterns and point an investigator to the origin of a fire. It can also help distinguish areas burnt by fires of separate origins after they have run together.

Reconstructing the pattern of a fire's spread through eyewitness observations or physical evidence alone can be difficult. However, in combination these two sources can provide sufficient information to establish how the fire spread and to estimate the speed at which it travelled. Even when investigating only the cause and origin of a fire, it is useful to consider information over the fire's entire life. This will help ensure that conclusions are consistent with observations elsewhere and with known patterns of fire behaviour.

Investigation of the physical evidence should commence as soon as possible after a fire. Interviews with eyewitnesses are better delayed for two or three weeks, to allow landholders to get on with the immediate tasks of stock disposal and cleaning up. After longer delays, though, an eyewitness's recollection may become vague or be influenced by discussion with others.

The investigator needs to draw together as much information as possible to explain how a fire starting at a particular location could logically develop under the prevailing weather conditions to burn-out the area involved.

Wind direction

The first thing to establish is the direction of the prevailing wind and any significant changes in wind direction during the fire. An anemogram from a location close to the fire will provide information on the mean direction of the wind during the fire's run, the extent to which it backed or veered over the course of the fire, and the direction and timing of significant wind changes. This will establish the direction of head fire travel at different times.

Most often, however, anemograms are not available from sites close to a fire. Even so, it is useful to pick up information from as many wind recording stations in the district as possible. This can be used to establish the general pattern of wind during the day and perhaps the rate of frontal movement across the countryside. These observations must be matched with conclusions on wind direction and wind changes in the fire area drawn from physical evidence left by the fire.

First, the overall shape of the fire should be mapped and any suppression activities noted. The major axis of the fire will provide the mean wind direction over the course of the fire, and the shape will help focus further investigations in specific areas. To learn about the extent of the fire near its origin or at the time of a major wind change, we need specific information on local wind directions in the fire area. This can be obtained from patterns of burning.

Lee-side charring on trees and posts

As the wind blows a fire past a tree or a post, the flames are drawn into the eddy zone on the leeward side and will extend further up on that side than on any other part of the tree or post (Fig. 9.1). The pattern of charring is often quite symmetrical and the peak point of charring will be directly opposite the direction of the wind striking the tree, indicating the direction the wind was blowing when that tree was burnt.

A large number of observations need to be taken into account to determine local variation in wind direction. It should be remembered that the direction of wind determined by this method is the direction the wind was blowing through the combustion zone, and may have been influenced by the fire's convection. Sometimes the width of the head fire and the position of either flank can be determined from differences in the direction of lee-side charring noted in a transect across the fire.

This technique is not always reliable. Evidence from split hardwood posts can be misleading, as rotting sapwood may burn away in preference to heartwood regardless of the post's orientation to the wind. Investigators should also look for evidence of the combustion of other fuels near the post that may have caused the post to burn on a side other than the lee side (Fig. 9.2). The best evidence comes from isolated trees or posts of uniform texture which do not continue to burn long after the fire has passed.

Grassfire investigation | 107

Figure 9.1: Flames in the lee-side eddy behind the trunks of two trees. Note how these extend to a greater height than the flames near the front of the trees. In many cases, the flames will leave sufficient charring to identify the direction of the wind past the trees.

Figure 9.2: Charring on a post caused by a timber gate burning against the post before collapsing. The prevailing wind direction was at right angles to the direction indicated by the char. The stockyard was heavily trampled and eaten-out, leaving only scattered grassy fuel which, when burnt, provided little physical evidence on the posts of the wind direction. Photo: J. Boath.

Leaf and stem freeze

When green leaves of shrubs or trees are scorched rapidly, they are 'frozen' in the position they were bent into by the wind when the fire passed through. The direction in which the leaves are bent indicates the direction the wind was blowing.

This technique is best used when shrubs and trees are in exposed positions and when fires have burnt during strong winds. Intense burning in a clump of shrubs may result in strong upward convection, which can freeze the leaves vertically. Many observations should be made, to cover local variations in wind direction.

Both leaf freeze and lee-side char indicate the direction of the wind through the flame zone when a fire passed a particular point. This evidence alone does not indicate whether the fire was travelling with or against the wind, or whether the wind direction at that point was that of the prevailing wind. However, it can be used with other evidence to build up a picture of the pattern of fire spread.

Partial burning

Very sparse pastures may be only partially burnt if the wind is very strong. This effect will show up in aerial photographs as narrow lines of burnt ground, sometimes not continuous, across eaten-out pastures and aligned in the direction of the wind (Fig. 9.3).

Shape of the fire perimeter

Ash patterns

In some pastures, differential burning in the lower layers of the fuel bed occurs as the flanks alternate from a backing to a heading fire. This leaves a pattern of ash of different colours (Fig. 9.4).

It is rare to find complete ellipses as illustrated in Figure 9.4. Under strong winds the head of a fire is always a heading fire, even during relative lulls in the wind. However, as the wind direction fluctuates the fire on each flank will alternate between backing and heading. Therefore an ash pattern left on one flank will have been created at a different time from a similar pattern on the other flank. Nevertheless, these ash patterns can be used to delineate the shape of a fire and, together with eyewitness reports, be used to build up a chronological map of fire spread.

Patterns in the ash may not be obvious from the ground but may stand out clearly when viewed from the air or in aerial photos taken within a day or two of the fire.

Other evidence of differential burning

Many indicators reflect the differences in intensity of backing and heading fires, and these can be used to help define the shape of a fire. In some grasslands, the difference between backing and heading fires may be apparent from differences in the

Figure 9.3: Aerial photograph of the Narraweena fire, 16 February 1983. The fire patterns show how the wind direction progressively changed after passage of the front at 3.45 p.m. (a) 3150: prevailing wind before 3.45 p.m. (b) 2800: 3.45–4 p.m. (c) 2450: 4–5 p.m. (d) 2250: after 5 p.m. Source: Dept of Lands, SA.

height of the remaining grass stalks. Backing fires tend to burn much closer to the ground, while the protective ash deposited by a heading fire leaves a length of grass stalk protruding above the soil surface even after the ash indicators have blown away.

In heathlands, the fluctuation in wind direction along the flanks causing alternate backing and heading fires can leave narrow strips of scorched shrubland with intact leaves. These indicate periods of backing fire spread; the heading fire completely defoliates and blackens the shrubs. This effect is most obvious in dense uniform fuels, and is commonly observed when fires burn through unthinned pine plantations. It is not often seen in the much less regular open eucalypt forests.

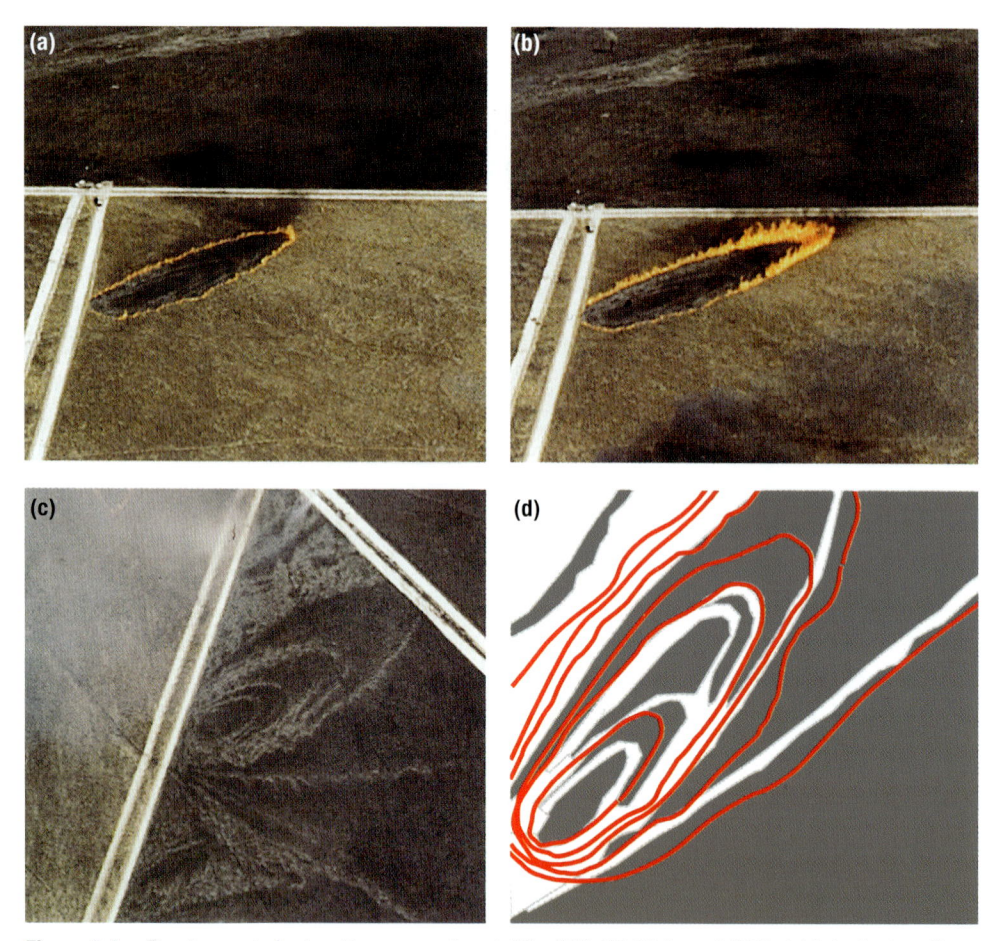

Figure 9.4: Development of ash patterns: experimental fire E26, NT, 20 August 1986. (a) 3.1 minutes after ignition. A lull in the wind speed has slowed the fire and around the entire fire perimeter the flames are drawn towards the burnt area. The resulting backing fire is depositing a white ash, which delineates the position of the fire perimeter at that time. (b) 3.6 minutes after ignition. The fire is spreading as a heading fire around 75% of the perimeter, including the left flank, and is depositing black ash. (c) Oblique aerial photograph of the ash pattern remaining after E26. (d) The measured spread pattern of E26 (red) superimposed on a map of the ash pattern left behind. This pattern allows an investigator to focus quite accurately on the origin of the fire.

The height of charring on trees and low shrubs gives an indication of flame height. When shrubs are not present, as is mostly the case in grasslands, the height of leaf scorch on scattered trees may help to indicate differences in the intensity of backing and heading fires. As much as possible, check that the fuel loads in the areas being compared were similar before the fire.

Eyewitness observations

Reconstruction of the spread of a fire from eyewitness observations is possible when there are a large number of witnesses. The reliability of each witness in estimating time and location should be noted. Rarely is any one witness completely accurate, but evidence from a large number of witnesses can establish the shape and rate of fire spread with reasonable accuracy.

At times, eyewitness reports can paint a very confusing picture of a fire, particularly when taken in isolation and before the probable pattern of spread has been established. It is necessary to appreciate that the witnesses may have been through traumatic experiences and are reporting on something they have probably never encountered before. Smoke probably restricted visibility, so investigators should try to establish how far a witness could see by asking questions about the smoke.

The most reliable data obtainable are the location of the observer, the time the fire reached a particular point and the orientation of the perimeter at that time. This involves physically taking the eyewitness back to where they were during the fire so that the investigator can understand exactly what they observed. The investigator should try to establish whether the observer could actually see flames from the location, or see only smoke. It is difficult to estimate how far away smoke or flames are unless it is possible to refer to the flames passing an identifiable feature such as a tree or windmill, or even a position on a slope relative to ridges and gullies. Across rolling topography, bushfire smoke always appears much closer than it really is.

Most observers can recall accurately the direction they were looking in, but again it is essential to go to the exact viewing point and record the bearing of the observation. Sometimes it is possible to obtain a cross-bearing from an observer at another location, for confirmation.

Establishing the time that the head fire crossed a road can often be a problem. Some observers may assume that the fire had just crossed the road because it was across it when they drove to that point. Investigators should try to establish whether the observers saw a head fire or a flank fire by asking questions about visibility along the road as they approached. If there was dense smoke across the road, it is quite likely that they observed the flank, or a tongue of fire parallel to the flank, approaching the road.

Establishing the time of the observation is very difficult. In general, people's impression of elapsed time during a fire tends to be shortened. Investigators should try to establish whether the witness had carried out some activity at a scheduled time that can be checked, e.g. listening to a particular radio program such as an hourly news bulletin, or meeting a school bus. If the place where the time was noted is different from that where the fire was seen, the investigator should try to reconstruct activity times to the point of observation of the fire. These are mostly longer than witnesses estimate.

Power failures may indicate the ignition time of a fire caused by a transmission line fault. Failures may also occur when high flames pass under a transmission line, causing it to short to ground through the ionized gases.

The investigator should ask for photographs and videos. Digital images with time data are extremely valuable but the clock in the camera needs to be checked for accuracy (and daylight saving time). The location from which the image was taken must also be checked to establish the position of the fire with reference to features in the landscape. The increasing prevalence of GPS (global positioning system) receivers in vehicles and mobile phones should provide unprecedented access to accurate location data. As with all sources of data, care should be taken to ensure the waypoints selected match those of the eyewitness observation.

Remote sensing tools such as infra-red and hyperspectral imagery from satellite or aerial platforms provide an additional source of fire behaviour data. However, a thorough understanding of how these systems work is necessary to provide correct interpretations of fire spread. This is particularly so for infra-red imagery, which returns a limited false-colour spectrum of temperature based on radiative energy received at the sensor. Depending on the dynamic range settings, infra-red imagery may show the heated air in and around the fire as the same colour as flame and thus obscure the real fire edge.

Determining the origin

Determining a fire's origin may be straightforward if firefighters arrived early and suppressed the rear perimeter of the fire soon after ignition. It may be very difficult if suppression at the back of the fire was delayed or burning-out operations were conducted upwind of the suspected ignition point. Single ignition points from backburning operations may leave a pattern identical to that of the original ignition.

Investigators should use all the indicators described above to establish the approximate location of a change from a backing fire to a heading fire in the direction of the prevailing wind. If the time when the very back of the fire was suppressed is known, and the probable ignition time is also known, it is possible to use the fact that fires backing directly into the wind spread at a constant rate (see Fig. 4.9) to locate the probable region of ignition. The elliptical shape of the back flanks can help focus the search for the ignition source, but the timing of lateral spread abreast or ahead of the ignition point is difficult to determine because of alternating heading and backing behaviour of the flank fire.

Power line ignitions

Power lines are a significant cause of wildfires on days of extreme fire danger. Fires may be caused by sparks from conductors clashing together, sparks from expulsion drop-out fuses, conductor breakage or large birds being blown across two conductors. A contentious point is the possibility of fires starting as a result of high-voltage conductors sagging under high temperatures and coming into contact with live trees.

When a conductor first comes in contact with the tips of the branchlets at the top of a tree growing under a power line, the electricity will short to ground through the tree, pruning off the green tips and killing the upper portion of the living twig.

Figure 9.5: The result of repeated electrical pruning of a tree growing under a high-voltage power line.

This process may continue over several years as the tree grows up, and has been described as electrical pruning or electrical deadening. After several years the top of the tree will show considerable concavity consistent with the conductor's sag and arc of sway on a hot summer day (Fig. 9.5). The twigs on the very top of this arc will show evidence of repeated resprouting, and their ends will be carbonised (Fig. 9.6).

The initial arcing to fine green material is unlikely to start a fire but, eventually, intermittent contact with the pruned and charred tips or the live branchlet below the arc-killed tip will loosen glowing embers of carbon or ignite flakes of dead bark or leaves in the tree top. These can fall to the ground and start a fire on a day of extreme fire danger.

Any tree that has developed the typical concave shape illustrated in Figure 9.5 is a potential source of ignition on days of extreme fire danger and should be cleared from beneath the power line.

Figure 9.6: A branchlet from the top of an electrically pruned tree near the origin of the Taylors Flat fire, NSW, 24 January 1975, showing (a) successive resprouting and (b) carbonisation of the twig end. Photos: J. Banks.

# 10
SAFETY: MYTHS, FACTS AND FALLACIES

People should never die in a grassfire. The heat pulse is transient and, although intense, lasts only a few seconds. However, during almost every serious grassfire season people die in situations that are all too common – and all too commonly unnecessary. These situations have happened before. They have been described in detail, both in training sessions for rural people and in the popular press. So why aren't we getting the message across? Bushfire lore is a fertile ground for myths and fallacies and it is our belief that some popular misconceptions about what happens in a bushfire may cloud the facts.

Survival in a grassfire is straightforward but there are three basic actions you must take.

- You must find an area that will not burn (at least for the duration of the passage of the fire front), and where you are protected from direct contact with the flames.
- You must protect yourself from the radiation from the fire.
- You must protect your airways from the heat, smoke and ashes produced by the fire.

A high-intensity grassfire can be a truly terrifying experience if you are directly caught in it. However, if you are not involved directly with the fire and view it only from a distance, a really serious grassfire often appears as just another large white cloud near the horizon. Most people in Australia are unfamiliar with wildfires of any sort. A grassfire travelling at 15 km/h or more, burning-out thousands of hectares of countryside and causing millions of dollars worth of damage, is a phenomenon they will be unlucky to experience even once in their lifetime.

Large grassfires produce a huge volume of smoke which, because our dangerous summer winds are north-westerly or westerly and fires occur in the afternoon, blocks out the sun. Directly downwind of an approaching large head fire it can be so dark that it may seem like night. Visibility may be reduced to 50 m or less, and effectively to zero as the head fire passes. The atmosphere turns red as light from the flames is reflected by the smoke. The noise of the fire and wind is extremely loud and can be deafening. Many people have described it as sounding like a steam train bearing down on them. These conditions, along with the dust, heat and high winds, are liable to set the imagination running wild. Inexperienced people may well be excused for exaggerating the phenomena they are confronted with, and offering explanations that defy the laws of physics.

To people who have never been in such a situation, exaggerated observations and wild explanations can sound plausible and can be difficult to refute. Even experienced people can have trouble refuting them. Such observations and explanations may become myths and fallacies, creating problems for fire control officers giving instruction on fire behaviour or expert evidence in legal proceedings. Such distortions have sometimes led lawyers to prefer the evidence of an eyewitness to that of an expert even though the described events may be clearly impossible. Most importantly, though, these myths and fallacies are seen as more believable than the cold hard facts and can prevent inexperienced people from taking proper action when confronted with a life-threatening situation.

Nevertheless, within most myths and fallacies there is usually a grain of truth. In this section we address the most common myths and fallacies, explain how they arose and how a proper understanding of fire behaviour can enable people not only to survive in life-threatening situations but to live safely in our flammable rural countryside.

The fire consumed all the oxygen around me and I could not breathe

For some reason, people involved in large fires appear to be more concerned about being deprived of oxygen or poisoned by carbon monoxide or another imagined toxic gas than they are about being burnt by the flames. We often hear comments such as, 'The fire used all the oxygen – I could not breathe.'

Flames require oxygen, and at oxygen levels of 13% or less flaming combustion cannot be supported. However, people can survive at these oxygen levels. So what happens? Certainly, if you were inside the actual flaming mass you would encounter uncombusted hydrocarbon gases and there would be a lack of oxygen. But this would not matter because you would already be dead – killed by the temperature and thermal effects of the fire.

If you are in a location where you are protected from the flames, there is always plenty of oxygen in the air being drawn into the fire. However, breathing difficulties can be created by the hot gases in the smoke and by the smoke itself. Smoke is a strong irritant; it can choke many people and make them feel starved for air. When the smoke is hot, the effect is magnified and can cause death.

Survival in the open in the front of a fast-moving grassfire is not easy. It doesn't matter much whether you suffer heat stroke from fierce radiation, asphyxiation from smoke and hot gases, or are burnt so badly that your immune system fails and you die from pneumonia months later – you end up dead. However, one thing is certain: if you are away from the peak heat pulse of the fire and protect your lungs from the smoke you won't suffer from lack of oxygen.

If you can shelter in a farmhouse or outbuilding you should not perish in a grassfire. This is your best option – people have left well-protected farm buildings and perished in their flight from the approaching fire (this will be discussed later). If you are on foot, on horseback or in an open vehicle there are certain actions you must take.

- You *must* find an area that won't burn – the bigger the better. This may be an area of burnt ground, a road, a swampy depression with green grass, a dam, creek or other waterhole, an area you burn-out and step onto from the upwind edge or any other area without combustible fuel.
- The next step is to protect yourself from radiation – again, this is not easy in the open. You have to lie flat on the ground and cover yourself as much as possible. A heavy woollen blanket or any additional heavy clothing could be useful. There may be a non-flammable obstacle you can place between yourself and the fire. In sandy soils, some people have survived quite severe fires by getting into a wheel rut and covering themselves with sand.
- The last and most important action is to protect your airways. This can be done by breathing into the ground, or covering your mouth and nose with a piece of wet material and breathing through that.

If you are overrun by a head fire, the flames will last only 10–15 seconds. In the flaming zone itself, the air temperature will be highest near the ground (see Fig. 10.1). After the flames have passed there will be a period during which the wind speed increases strongly (see later myth 'The fire creates its own wind') and carries a lot of ash, sparks and smouldering debris at ground level. This may last one to two minutes so you need to protect your airways not only from hot gases while the flames are around you but also from the smoke, sparks and blown debris after the flames. Stand up as soon as possible after the heat of the flames has passed, to get

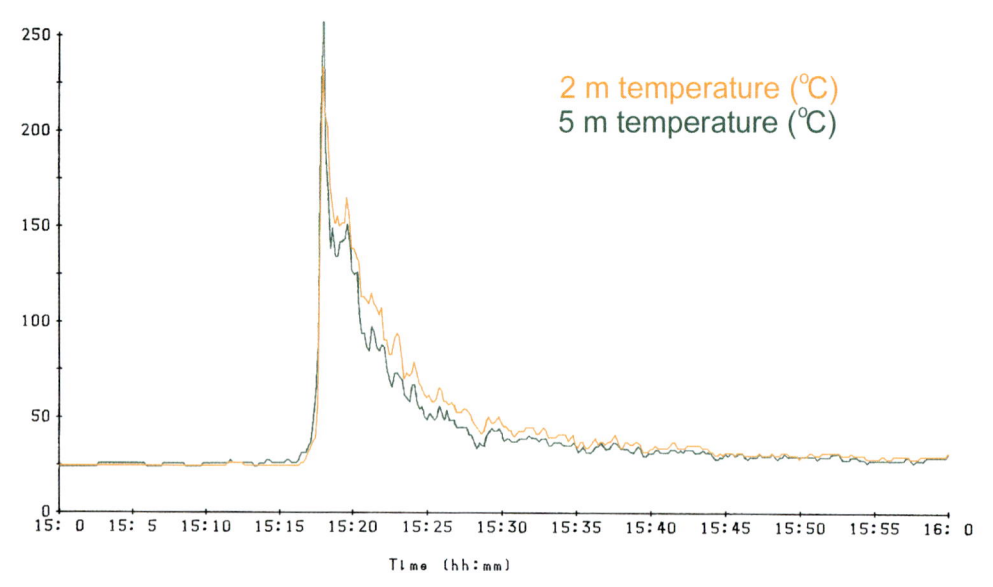

118 | Grassfires: Fuel, Weather and Fire Behaviour

Figure 10.1: Air temperatures in a crowning forest fire measured at 2 m and 5 m in a 10 m-diameter clearing. In a grassfire, the magnitude and duration of elevated air temperature will be less. Source: CSIRO Project Aquarius (unpublished data).

above the smoke and hot air at ground level. The air behind the fire will be cooler and cleaner at 1.5 m above the ground.

You need to be fully clothed – long trousers and long-sleeved shirt – to protect yourself from radiant heat and burning debris after the fire has passed.

There is no easy or failsafe advice if you are caught in continuous grassy fuels without access to any sort of clearing. There have been cases, during prescribed burning of shrubby forest, where people have run through the flames of the head fire (albeit under relatively mild conditions) and survived on the burnt-out ground, where others who sought to shelter on the unburnt ground perished.

The flames at the head of a fast-moving head fire may be 10–50 m deep and up to 8 m high, and in our opinion it is highly probable that the thermal pulse from these flames will kill you before you have travelled a couple of metres. The depth of flames on the flank is always narrower and you may be able to run onto burnt ground through flames that are only a couple of metres deep. Moderately fit people can hold their breath while running for more than 10 seconds and cover more than 50 m. This will make your exposure to flames, radiant heat and dense smoke shorter than if you lie down in the fuel, which would certainly kill you even on the flank. However, everything has to go right for this to be successful. You need a deep breath of cool clean air, the depth of flame has to be short and you cannot fall or stumble.

On the flank of a fire it is far better to wait for a switch in wind direction that changes the flank to a backing fire then run across the lowest flames. You may still have to run some distance through smoky smouldering fuel to get onto ground that was burnt five or more minutes earlier. The air over this ground should be relatively clear. There may even be a strong lull in wind strength which reduces the flames at the head of the fire. However, using this technique requires a fair bit of knowledge about grassland fire behaviour and a good knowledge of your position in relation to the fire as a whole.

We consider it essential that you get onto burnt ground and seek protection from the flames, however flimsy that protection may seem – when push comes to shove it is always better to seek the oven than the griller.

A case story

On Ash Wednesday 1983 in South Australia, a farmer was caught in the open on a tractor towing a water trailer when a fire broke away from a 35 km flank after the wind changed. He was in an area of swampy ground with some green patches. Visibility immediately fell to close to zero, and spotfires started to appear around him. As he attempted to drive away from the fire and get onto a large area of moist green ground, the trailer-tank overturned and stalled the tractor. The farmer had no option except to climb into the empty tank. He said later, 'I thought I had lost my dog who was out there with me but when I got into the tank there he was inside. He had beaten me to it.' Climbing into the empty overturned water tanker saved both their lives.

Boiled alive

Most texts and pamphlets on bushfire survival include a warning to avoid sheltering in elevated watertanks because immersion in water above body temperature prevents you from cooling yourself by sweating. This results in a rapid rise in core body temperature to a point of physical collapse (immersion in water at 45°C will result in loss of consciousness in about three minutes). There were reports of people perishing in elevated watertanks during the 1939 fires but there is no information about the location of those tanks and their proximity to burning buildings or other sources of prolonged heating. This is unfortunate, because it seems to have led to the avoidance of watertanks in general.

If a tank contains a substantial amount of water it is unlikely that the water temperature will be above the mean daily air temperature. It takes a considerable amount of heat to raise the temperature of this water by 1–2°C. The passage of a grassfire is unlikely to raise the temperature of a watertank by even a fraction of 1°C, so any tank or water body out in the open is an excellent place to seek refuge for a few minutes, provided you can get out afterward.

If you are in a dam or other water body you need to pull wet clothing over your head to make an air pocket and protect your airways from hot gases and blown ash until after the fire has passed. A cattle watering trough may be an excellent place to take refuge, particularly if the grass has been eaten-out and trampled around it. The banks of a narrow creek may contain a substantial amount of fuel, particularly if it has been fenced off from grazing. However, a waterhole in the creek may be the only place to survive and if you continue wetting the clothing over your head you can easily survive until the vegetation burns out completely. During the Como/Janalli fire in January 1994 three people survived in a backyard swimming pool on the lee-side of the house while the house burnt down next to them. They were in the water for more than an hour and protected their airways from the horrendous heat of the burning house by continually wetting towels placed over their head. This is a far longer period of exposure than is ever likely to be experienced in natural fuels, particularly grass.

Of course there are dangers. If you can't swim but you jump into a deep steep-sided tank you will probably drown. If you need a ladder to get in, take the ladder in with you to get out! By and large, shallow water and moist areas are refuges that definitely will not burn and if you have to survive the passage of a grassfire you certainly will not be boiled alive.

The fire creates its own wind

One of the most common statements about bushfires is, 'The fire creates its own wind and you can't do anything about it after that!' The problem is not the initial proposition that the fire creates a wind field around it – it does – but the idea that the in-draught winds induced by the fire keep driving it forward independently of the prevailing weather conditions.

Fires in heavy fuels such as forest slash from clearing operations generate very powerful convection currents, which at times may form fire whirlwinds or fire tornadoes, because of the high rate at which heat is released by the fire. These currents can induce strong in-draught winds that are drawn into the fire zone from all around. When the prevailing wind is <15 km/h, this convection will dominate the prevailing wind and the fire is actually contained by its own in-draughts. In grassfires, however, the total fuel load is relatively light and the in-draught winds will contain the fire only when the prevailing wind is light (<5 km/h).

In forward-spreading fires, the convection column is inclined by the prevailing wind. The main in-draught is into the back of the flames behind the fire front. This in-draught increases the speed of the prevailing wind behind the fire and is drawn downward into the combustion zone before being carried up into the convection column. The convection column effectively blocks the prevailing wind, creating a zone of light and variable wind for some distance ahead of the fire front (Fig. 10.2). The increase in the prevailing wind does not extend more than 50–100 m beyond the perimeter of the fire, although the decrease in wind speed may extend several hundred metres downwind of an intense fire.

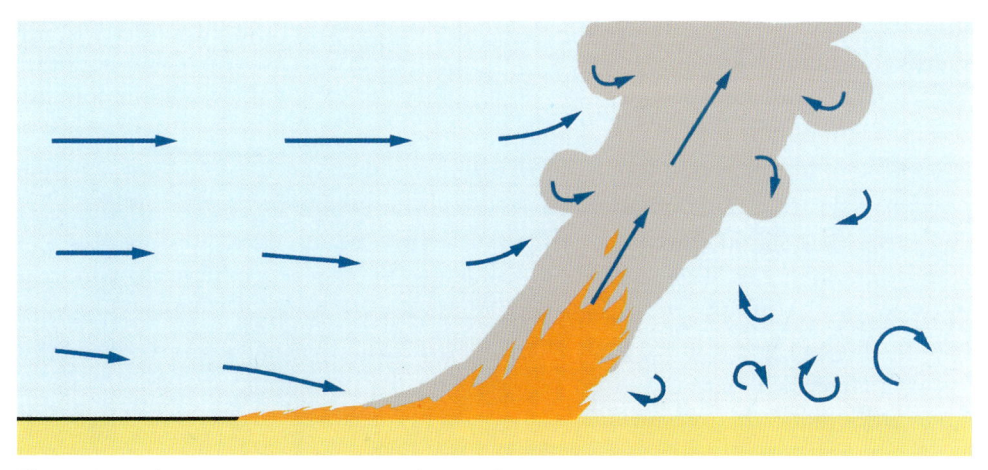

Figure 10.2: Cross-section through a head fire. The fire-enhanced wind flows into the back of the fire. There is little in-draught on the downwind edge of the flames, but the convection column blocks the prevailing wind and creates a zone of light and variable wind ahead of the fire.

During experiments with grassfires in light fuels, it was possible to walk about 20 m behind the flames in a zone of clear air as the in-draught was pulling smoke-free air down into the head fire. This structure can be readily observed when a fire passes over a firebreak; the enhanced wind behind the flame front flattens the flames after they hit the break (see Fig. 8.8). This wind may be strong enough to move smouldering debris on the ground.

As discussed in Chapter 6, the behaviour of a fire is strongly influenced by the eddy structure of the atmosphere and by gusts and lulls in the prevailing wind. When the wind speed drops during a lull, the enhanced wind behind the fire front does not keep propelling the fire forward. Rather, the convection column straightens and its base may move back onto the burnt area, creating an in-draught on the downwind side of the fire front. This dramatically slows the fire's spread.

One situation where a fire appears to create a strong driving wind is when it starts at the bottom of a steep slope under still conditions. The fire's thermal draught may enhance light upslope thermals and induce a wind in the vicinity of the fire which will propel the flames up the slope. Once the fire reaches the top of the slope the thermal convection is reduced and the surface wind dies away. In this case, the driving in-draught wind is a natural consequence of fire burning on a steep slope.

Another situation is the strong circulation that may be associated with a number of high-intensity forest fires burning in close proximity at the same time. During the 2003 ACT fires two intense forest fires were located 7 km apart. A breakaway from the rear of the northern fire occurred and travelled between the flanks of the two fires. The convection from the well-established forest fires towered to over 8000 m. It apparently created very strong winds between the two flanks which drove the breakaway fire at around 20 km/h through eucalypt forest, pine plantation and sparse grassland about four times faster than the main fires on either side.

Shortly before the breakaway burnt out of the forest, the combined interacting convection columns had nearly doubled in height to 14 000 m and created a fire whirlwind of near-tornado intensity. This whirlwind travelled over 14 km, creating a swath of destruction about 200 m wide with rotational wind speeds estimated at 200 km/h. It proceeded ahead of the surface fire and appeared to drag that fire along. Although this was certainly a fire driven by convection-induced winds, it burnt mainly within the fire complex and was powered mostly by convection from fires burning in heavy forest fuel. The whirlwind and the strong winds associated with it petered out after it crossed into the suburban area and the intensity of the surface fire reduced.

Extrapolation of relationships derived from fires of low intensity to fires of extreme intensity may help support myths about fire and wind. For example, the nomograms in Figures 5.10 and 5.11 show that in extreme conditions a fire that travels at 20 km/h on level ground under a wind speed of 45 km/h will attain a speed of 80 km/h up a slope of 20°.

A fire will certainly travel rapidly up a steep slope but it is doubtful that its speed would be more than twice that of the prevailing wind. A number of factors come into play. First, the wind speed on the slope will be higher than that measured on level ground, as explained in Chapter 6. Second, under high wind speeds the convection column from the fire may not lift away from the slope (Fig. 10.3). In this instance, in-draught winds will blow the flames directly into the fuels on the slope, causing very high rates of spread.

While there is no doubt that fires can sometimes induce quite violent winds near the ground, the speed of the fire over the ground is determined by the prevailing wind and the topography.

The fire was so hot it melted the engine block

We see this claim almost every time there is a major fire, when a reporter stands in the ashes of a burnt-out house or examines a burnt-out vehicle. The implication, either explicit or indirect, is that the flames of the wildfire were so hot that they melted an engine block and therefore it would be impossible for anyone to survive in the vicinity. This fallacy brings us into the difficult area of measurement of flame temperature and heat transfer. It also brings us to an explanation of a colleague's statement that he 'only ever measured flame temperature to the nearest half-brick'.

As discussed in Chapter 3, while the diffusion flames of a grassfire may appear solid and continuous they are in fact the result of the complex turbulent folding of unburnt hydrocarbon gases (fuel), air, the narrow reaction zone between the fuel and the air, and combustion products around the reaction zone (Fig. 3.2). The temperature within the reaction zone is around 1600°C and is measured with special

Figure 10.3: Under very strong winds, the convection column will not lift away from the surface of steep slopes. Flames will be blown directly into unburnt fuel, resulting in very high rates of spread.

instruments called optical pyrometers. Because the reaction zone is folded due to the turbulence of the wind and combustion effects, the mean temperature of an entire flame is much less. Thermal measurements of flame fronts in grassfires have found that the temperature of a flame ranges from around 300°C at the tip to around 1100°C at the base if the flames are more than 6 m high.

The temperature of something embedded in the flames depends on the transfer of heat from the flames to the object. It thus depends on many factors such as the duration of heating, the mass, thermal conductivity and the emissivity of the object. In a grassfire that remains at a point only for 5–10 seconds, there is little time to transfer a lot of heat to a solid object, whereas the prolonged combustion of a house or vehicle is sufficient to raise the temperature of quite a substantial lump of metal to its melting point. Our colleague was more interested in the temperature of objects immersed in flames (the half-brick) than in the flames themselves.

When there is sufficient heat to raise the flammable materials in a car to ignition they can burn rapidly and hotly. The problem of surviving in motor vehicles is discussed below.

The green grass burns faster

One piece of folklore is that fires burning in pastures not quite fully cured will burn faster than those in fully cured pastures. That is, fires burning in fuels 80–95% cured appear to spread faster than those in fuels 100% cured. While this observation may seem contrary to our curing function in Figure 4.7, there is a simple explanation. Once grasses are fully cured, they start to break down. After some time the grass will suffer mechanical damage due to wind, rain or grazing from native animals; this reduces the height of the grass and increases the fuel's compactness.

Figure 10.4: A woman was killed in this car. She was travelling down a narrow lane in the south-east of South Australia during the Ash Wednesday fires and crashed in dense smoke into heavy fuels on the roadside verge. The fire in the open paddocks was less intense than the fire beside the road.

This process will be accelerated if the pasture is grazed by stock. Flame heights in standing grass are always higher than flames in eaten-down pasture (see Fig. 4.5) and give the appearance of spreading faster although in fact there is little difference in rates of spread.

The relationships between wind speed and fire spread (Fig. 5.7) show that a fire will spread about 20% more slowly in a grazed pasture, and 60% more slowly in an eaten-out pasture, than in tall standing grass at any wind speed. A fire in an eaten-out pasture that is fully cured will spread less rapidly than a fire in an ungrazed pasture that is partially cured, e.g. 85%.

Another factor that may promote the illusion that a partially cured pasture burns more rapidly than fully cured grass is the additional noise and smoke usually associated with the combustion of grasses that have some green material mixed with them.

It seems unlikely that this myth has led to people avoiding safe areas of green grass but it is worth remembering that there are often green non-flammable grasses associated with wet soaks in pastureland. During the Narraweena fire several people who were killed on the roadside were quite close to unburnt green patches where they could have survived.

Exploding petrol tanks

A fear that many people still hold is that if they stay in their car when trapped by a grassfire the petrol tank will explode. This fear may have been reinforced by movies, where almost every car exposed to any sort of flame explodes with a bone-shaking blast (ably assisted by several charges of explosive).

Figure 10.5: The driver of this car was killed after he abandoned his car ahead of the fire. Although the car eventually burnt out (probably starting at the rear tyres) the car would have provided good protection during the passage of the fire front.

In reality, petrol tanks don't explode. When they are subjected to strong heat, the petrol vaporises and expels all the air in the top of the tank, making the mixture too rich for an explosion to occur. The vapour may be blown out of the filler cap or vent, and burn strongly when it comes in contact with the air.

It is really quite difficult to get a petrol tank to explode. An explosion requires the exact mixture of 14% petrol vapour to air – as produced by a carburettor or fuel injector to make an engine run. In a petrol tank, this mixture can be present only if there are a few millilitres of petrol in an otherwise empty tank. For example, petrol tanks have exploded when people have tried to weld them without first flushing them thoroughly with water – the residual vapour is sufficient to form the explosive mixture.

In most situations, a normally maintained car will provide good protection from a grassfire even if the car subsequently catches alight and burns out. The time taken for a car to burn-out is quite long. Fire usually progresses from the tyres through to the engine bay, into the oils and greases around the engine, into the linings of the car and last to the petrol tank. The heat of the burning car will melt or burn some part of the fuel line, draining the petrol underneath the car. This will indeed burn quite violently. However, all this takes time. In a grassfire, the grass around the car will burn in less than a minute and even if the car does catch alight you will have time to get out of it onto the burnt-out ground.

It is always best to park the car where there is the least fuel. In some farmland, fuel along the roadside is much heavier than elsewhere. Not only has there been no grazing, but litter and branches from the trees will have contributed to the fuel load. As a result, fire along a roadside may be more intense and more persistent than fire in a pure grassland fuel. In the 2005 Wangary fire a vehicle that ran off a gravel road into roadside fuel burnt and the occupants died. An adjacent vehicle parked on the

Figure 10.6: This fire truck was hung up on a log. By the time the photograph was taken the tanker was well alight. The occupants walked to a nearby road before the fire came, but could have survived in the cabin even though tyres, hoses and petrol were burning on the back of the truck. Photo: J. Cutting.

road suffered minor damage but protected the occupant. In some cases, it is safer to drive into an eaten-out paddock away from the road to find an area where grass is sparse and the fuel is light.

These steps will increase your chances of coming through safely:

- Park the car on bare ground or where there is least fuel surrounding the vehicle.
- If possible, park the car into the wind so that, if the fuel tank catches alight, the wind will blow the flames away from the cabin (see Fig. 10.6).
- Turn on the headlights and hazard lights, particularly if you are parked on a road, and make sure the windows are wound up tightly.
- If you have driven into flames, do not try to turn around. You will probably get stuck in a roadside drain or heavy roadside fuel.
- Try to watch what is happening outside, although as the flames arrive you may need to get beneath window level briefly to protect yourself from radiant heat coming through the windows.
- If possible, cover your nose and mouth with a wet cloth – gases from the burning paint, door seals or even the fire itself could enter the car shortly after the flames hit.

After the flames pass there will be a minute or so when blown ash and smouldering debris immediately behind the flame front will completely obscure your vision outside. Be confident – the air will rapidly clear and you can step out of the car onto the burnt ground.

Cars will burn, as discussed above. We are confident that a car will provide protection during the passage of the hottest grassfire, but it may not be suitable protection against a forest fire or even fires burning in heavy fuels beneath trees lining a narrow road. Trials to simulate the impact of forest fires showed that, while a vehicle could withstand considerable heat, very high-intensity fires could result in catastrophic failure. Exposure to a simulated crown fire caused linings to smoulder, filling the car with dense acrid smoke after 8 seconds. Failure of the windscreen after 20 seconds led to influx of flame and flashover of the lining material 8 seconds later, filling the whole cabin with flame and molten plastics. Survival in this situation would be impossible and the only advice under extreme conditions is don't get caught in the forest.

Resist the temptation to flee from a farmhouse in the face of fire. You are always much safer at the homestead, sitting out the fire, than fleeing along smoke-filled roads. Numerous tragedies have occurred when families have left their houses in the face of a fire only to crash and be killed on tree-lined roads. Firefighters who have arrived at the scene shortly after the fire passed have found that the houses were either intact or burning very slowly. In either case they would have provided a safe refuge. If you stay with the house, not only will you survive but you also have an excellent chance of saving the homestead by putting out any small fires lit by blown embers.

Great balls of fire

A popular expression, particularly in the media, is that a homestead or town was consumed in a fireball. The term is used loosely. Inexperienced observers have often described their experience in terms of fireballs, often of huge dimensions, advancing well ahead of a fire front and suddenly igniting everything around them. People watching forest fires from a distance often get the impression that there are balls of fire rolling along the tops of the trees, and some have offered the theory that these are supported by volatile gases brought out of leaves by the hot summer sun. It is also suggested that these gases carry the fire across the tree-tops without a supporting surface fire beneath. Another suggestion is that fires can produce a huge ball of combustible gas akin to a boiling liquid expanding vapour explosion (BLEVE) rolling across the countryside, making survival impossible even on bare ground.

Separate envelopes of burning gas can detach from surface flames. However, the gas is produced by the combustion of surface fuels and, once detached, these envelopes burn up rapidly and cannot roll any distance downwind ahead of the flame front. In most cases, the flames of high-intensity grassfires are less than 5 m high, although brief flashes of flame may extend considerably higher. In forest fires the flames may be two to three times the height of the tree canopy, and on occasions

Figure 10.7: The Narraweena fire approaching a house near Penola at around 4 p.m. (a) Light reflected in the smoke gives the appearance of massive flames. (b) The flames in the grass are actually less than 1 m high. (c) The flame front has passed and burnt out the fine fuels and fence palings of the fence. (d) The air behind the front is clear. Large material is still burning. Photos: R. Jackson.

towering columns of flame have been observed rising at an angle of 60–70° and extending 300 m above the ground. These towering flames are short-lived, and result from the concentration of unburnt gas caused by very high rates of combustion.

When considering stories of these phenomena, it is worth remembering that visibility at the front of a large fire may be very low due to the smoke and it is often very dark, even in the middle of the afternoon (Fig. 10.7). In the darkness the light of the flames will be reflected back from the smoke, much as at night, and rolling billows of red smoke can give the impression that there are balls of flame higher up in the convection column. This can be a truly awesome spectacle. However, the height of the flames is determined by the amount and distribution of fuel, and extremely long and persistent flames cannot be formed unless there is abundant fuel to produce the gas. Any ball of unburnt gas will be rapidly diluted by the wind.

Faster than a speeding bullet

Well, at least faster than a beat-up Kingswood on a country road. After every wildfire there are reports of fires spreading at phenomenal rates. People describe their experience of driving down the road at 80–90 km/h and not being able to keep pace with the fire. On later analysis, when information has been drawn from all observers, the pattern of fire spread often reveals a predictable rate of 16–20 km/h for the maximum fire speed. What explains this convincing illusion?

Grassfires burn in a direct line across the countryside, determined by the wind direction. This is usually at an angle to the road network, which is often set out on a square grid. A person driving down the flank of the fire is likely to be driving sometimes towards and sometimes away from the fire front. The driver may have to negotiate right-angled bends and slow down for creek crossings and other obstacles, so the distance travelled is likely to be considerably longer than the path taken by the head of the fire. Considerable speed is required to catch up with the head.

Another feature of fire behaviour that can give the illusion of high rates of spread is waves of flames in the flank fires. As the wind oscillates, its change in direction progressively picks up the flames on the flank, giving the appearance of a wave of flame actively travelling down the edge of the fire. This wave will appear to travel at the speed of the gust, perhaps as fast as 80 km/h. Also, any switch in wind direction which hits the back of the fire first then travels down the flank will pick up the flames at the rate that the change in wind direction hits the flank. This can make it appear that the flames themselves are travelling very rapidly.

When first arriving at a fire, it is a good idea to take a few minutes to observe its behaviour and note the movement of the flames compared with the speed at which the flame front is actually travelling. It will soon become clear that, while a rate of spread of 18–20 km/h is certainly devastatingly fast, no miraculous rates of spread vastly exceeding that figure are maintained for any length of time.

The fire was equivalent to eight Hiroshimas

The dramatic comparison of the energy released by a bushfire and the energy of the nuclear explosion at Hiroshima, Japan, during World War II has been used by embattled emergency services personnel trying to hide their incompetence by giving the impression that there was nothing they could have done, by foresters trying to argue a spectacular case for the benefits of fuel reduction, and by reporters simply trying to grab a headline.

This is a classic fallacy that arises because of confusion between *total* heat release and *rate* of heat release. Many comparisons can be made between the total heat release of a fire and the heat released by oxidation of other objects, although 'It was equivalent to all the rusting cars in western Sydney' doesn't have quite the same ring. We will discuss everything in the same units and see how they compare.

The energy yield of Little Boy, the nuclear weapon dropped on Hiroshima, is estimated to have been equivalent to around 15 kilotons of TNT or around 63×10^{12} joules (a result of only about 1.4% of the uranium 235 in the bomb actually fissioning). The energy yield of a kilogram of grass is about 16×10^6 joules, so at an average

Figure 10.8: At times the convection column over a large fire looks like the typical mushroom cloud of a nuclear explosion but that's where the comparison ends. In this spectacular cloud over the Perth hills most of the energy release comes from the condensation of water vapour. The burning conditions at ground level are relatively mild. Photo: Li Shu.

fuel load of 4 t/ha a hectare of grass has a total energy yield of 64×10^9 joules. Therefore, Little Boy had the same energy yield as roughly 1000 ha of grass.

When it comes to doing damage, the total energy spread over a large area does not mean much – it is the rate at which that energy is released that matters. If we express our energy release in terms of energy release per second per area ($J/sec/m^2$ or W/m^2) we find that the rate of energy release of our rusting cars is negligible; that our grassfire travelling at 20 km/h through a fuel of 4 t/ha with a burn-out time of 15 seconds releases energy at around 0.4 MW/m^2; and that if we assume Little Boy released all its 63×10^{12} joules of energy through 0.1 m^2 at the source in a microsecond, this would result in an energy release rate of 64×10^{13} MW/m^2 – over a trillion times greater than our grassfire.

We shouldn't have to do the calculations to see the absurdity of this comparison. The bombing of Hiroshima killed 70 000–80 000 people in the first instant and around 140 000 people in the first six months. Even in the severest bushfire people can take action to seek protection. Comparing a bushfire to even a small nuclear explosion is an insult to the people of Hiroshima and Nagasaki and trivialises the seriousness of nuclear warfare.

It can't happen here

We often hear of property owners who are reluctant to take precautions against fire on their properties, justifying their stance by saying that 'It can't happen here', that other areas are in a 'natural fire path' or that a particular feature of their landscape makes it highly improbable, if not impossible, for a high-intensity fire to burn them out. It is true that in most areas of south-eastern Australia the likelihood of a high-intensity wildfire burning across a particular farm or property is quite low – perhaps once in 50 years. But the fact is that severe grassfires can occur anywhere in the agricultural belts of south-eastern and south-western Australia. All that is required is a good season with abundant grassy fuel that cures after a brief summer drought, a day of extreme fire danger with strong hot dry winds and an ignition source. The path of the fire is determined only by the location of the ignition point and the direction of the prevailing wind.

Better awareness, more stringent regulation and more efficient suppression during fire danger periods have dramatically reduced the number of fires burning in the countryside on days of less than extreme fire danger. There is, however, little evidence to suggest that the number of ignitions on the extreme days has decreased. There is always the possibility of accidents in the course of normal activities starting fires that, under less severe conditions, would either not start or would be easily controlled. There also appears to be more deliberate incendiarism than was the case in the past. The result is the same, and individuals throughout rural Australia should take basic sensible precautions to protect their major assets – their homestead and stock – from wildfire.

The red steer

The popular description of a fire as a 'red steer' – a raging out-of-control animal that escapes confinement and charges headlong across the countryside causing random destruction – is not really appropriate. The direction of a fire's spread is predictable, we can make a good estimate of how fast it will travel and we can forecast what it will do provided we have good information about the weather and weather changes.

Perhaps more importantly, the analogy is inappropriate when considering the damage that a fire will do. Fire is not a force over which landholders have no control. They can influence its intensity and the damage it does by determining the amount of fuel available for it to burn. Because farmers and graziers depend on grass for their livelihood or have crops such as wheat or plantation trees, there will always be situations where it is inappropriate or not feasible to reduce fuels. However, it is important that they reduce the amount of fuel around valuable assets that require protection. The fuel available for combustion – to revert to the 'red steer' analogy – is what feeds the 'steer'; if we take away its food the 'steer' will die.

Landowners are responsible for the fire intensity on their property – nobody else. Homesteads can be protected by ensuring that home paddocks are eaten-out and by keeping the surroundings of the house clean, with minimum accumulation of nearby flammable material. It is astounding how much flammable material some

Figure 10.9: Majura fire, ACT, 3 March 1985. Fire emerging from a pine plantation into grassland under conditions of extreme fire danger (GFDI 70). (a) The fire in the pines is clearly life-threatening. (b) The fire in the grassland, taken about a minute later, presents a relatively smaller threat. Photos: J. Cutting.

people have in the vicinity of their homes – nice native gardens mulched with woodchips, heaps of old tyres, old cars, stacks of sawn timber that will be used one day (don't know when), dirty windbreaks that have been allowed to build up material at ground level, wood heaps etc. All these things can contribute to the intensity and duration of the fire near the house and be an important source of firebrands.

Woodchip-mulched gardens have replaced green lawn in many places. During the 2003 ACT fires these gardens continued to burn, producing showers of firebrands for more than 20 minutes after the front had passed and contributing to the extensive loss of houses in the suburbs.

Woodlots and farm forestry operations, which are increasingly popular, are difficult to protect. Early in the life of a woodlot, farmers can lose their whole investment in a single fire. Even here, though, measures can be taken to reduce fuels. These include pasture improvement (preferably before planting) to remove unpalatable tussocks that build up large amounts of dead material, removing branch debris by slashing or careful burning under mild conditions when the trees are resistant to low levels of heat, and heavy grazing to remove palatable grass before summer and to trample other material.

Whatever the cause of a fire, its intensity is determined by the amount of fuel the landholder chooses to retain on the property. The landholder effectively owns the fuel and so determines whether the fire can spread and how intense it will be. In other words, the landholder owns the fire. How much damage a fire might do is a business decision for the landholder – who really cannot put all the blame on someone else!

Glossary

Adiabatic Thermodynamic change in state in which no heat is gained or lost by a system; compression results in warming, expansion results in cooling.

Advection The transfer of atmospheric properties by movement of air.

Air mass An extensive body of air having similar properties of temperature and moisture in a horizontal plane.

Ambient air Air of the surrounding environment.

Anchor point An advantageous location, generally a fire barrier, from which to start construction of a fireline. Used to minimise the chance of firefighters being outflanked by the fire while the line is being constructed.

Atmospheric stability The degree to which the atmosphere resists turbulence and vertical motion.

Available fuel The portion of the total fuel that would actually burn under specific conditions.

Back of a fire That portion of a fire spreading directly into the wind. Usually, but not always, that portion of a fire edge opposite the head fire and the slowest-spreading portion of the perimeter.

Back-burn A fire set along the inner edge of a control line to consume the fuel in the path of the head fire.

Backing fire A prescribed fire or wildfire burning into or against the wind or downslope without the aid of wind; characterised by flames leaning over the burnt area. See also **Heading fire**.

Backing wind A wind that shifts direction counter-clockwise.

Black out To systematically work the entire area of a fire for a defined distance in from the control line to ensure it is completely free of burning material.

Breakaway	(1) A fire edge that crosses a control line intended to confine the fire. (2) The resultant fire.
Burn	(1) An area over which fire has run. (2) A fire set deliberately to meet a management objective, e.g. back-burn.
Burning-off	Generally, setting fire (with more or less regulation) to areas carrying unwanted vegetation such as rough grass, slash and other fuels.
Burning-out	Setting fire so as to consume unburnt fuel inside the control line. See also **Back-burn**.
Burn-out time	The time taken for a particular unit of fuel to combust completely; includes both flaming and smouldering combustion. See also **Residence time**.
Bushfire	Any uncontrolled fire burning in forest, scrub or grassland.
Combustion	Consumption of fuels by oxidation, evolving heat and generally flame and/or incandescence.
Control line	A comprehensive term for all the constructed or natural fire barriers and treated fire edges used to control a fire.
Convection	(1) The transfer of heat by movement of gas or liquid. (2) In meteorology, atmospheric motions that are predominantly vertical in the absence of wind (which distinguishes this process from **Advection**).
Convection column	Thermally produced ascending column of gases, smoke, ash particulates and other debris produced by a fire. The column has a strong vertical component indicating that buoyant forces obstruct the ambient surface wind and can form a **wake zone** in the lee of the column.
Curing	The progressive senescence and drying out of a grass after flowering (annuals) or in response to drought (perennials).
Curing state	The fraction of dead material in the grass sward.
Dead fuels	Fuels having no living tissue and in which the moisture content is governed almost entirely by atmospheric moisture (relative humidity and precipitation), soil moisture, air temperature and solar radiation.

Direct attack	Any treatment of burning fuel, e.g. by wetting, smothering or chemically quenching the fire, or by physically separating the burning from unburnt fuel.
Eddy	Any circulation drawing its energy from a flow of much larger scale and brought about by pressure irregularities, e.g. in the lee of a solid obstacle.
Extreme fire behaviour	Extreme implies a level of wildfire behaviour that ordinarily precludes methods of direct control action. One or more of the following is usually involved: high rates of spread, prolific spotting, presence of fire whirls, a strong convection column.
Extreme fire danger	The highest fire danger class.
Fingers of fire	The long narrow extensions of a fire projecting from the main body.
Fire behaviour	The manner in which a fire reacts to the variables of fuel, weather and topography.
Firebrand	Any burning material originating from one fire that could start another fire, commonly bark but also leaves, seed heads, embers, sparks etc.
Firebreak	Any natural or constructed discontinuity in a fuel bed utilised to segregate, stop and control the spread of a fire, or to provide a **control line** from which to suppress a fire.
Fire danger	Sum of constant and variable fire danger factors affecting the inception, spread and resistance to control of a bushfire.
Fire danger index	A relative number denoting an evaluation of rate of spread or suppression difficulty for specific combinations of fuel, fuel moisture content and wind speed.
Fire danger rating	A fire management system that integrates the effects of selected fire danger factors into one or more qualitative or numerical indices of current protection needs.
Firefighter	A person whose principal function is fire suppression.
Fire intensity (Byram's fireline intensity)	The product of the available heat of combustion per unit area of ground and the rate of forward spread of the fire. The primary unit is kilowatts per meter of fire edge.

Fireline	(1) A loose term for any cleared strip used in the control of a fire. (2) That portion of a control line from which all flammable materials have been removed (by scraping or digging down to the mineral soil) or have been coated with fire-retarding chemicals. (3) A line cleared around a fire, generally following its edge, to prevent further spread of the fire and effectively controlling it.
Fire presuppression	Activities undertaken prior to fire occurrence to help ensure more effective fire suppression. Includes overall planning, recruitment and training of personnel, procurement and maintenance of firefighting equipment and supplies, fuel treatment and the creation, maintenance and improvement of a system of firebreaks, roads, water sources and control lines and pre-attack planning.
Fire risk	The chance of fire starting, as determined by the presence and activity of causative agents.
Fire scar	(1) A healing or healed-over injury caused or aggravated by fire on a woody plant. (2) The destructive mark left on the landscape by fire.
Fire season	The period(s) of the year during which fires are likely to occur, spread and do damage sufficient to warrant organised fire control.
Fire suppression	All the work and activities connected with fire-extinguishing operations, beginning with discovery and continuing until the fire is completely extinguished.
Fire threat	Sum of all factors that affect the inception, spread and difficulty of control of a fire and the damage it may cause.
Fire trail	A cleared way, broad enough to permit single-lane vehicular access in a remote area.
Fire whirl	A spinning vortex column of hot air and gases rising from a fire and carrying aloft smoke, debris and flame. Fire whirls range in size and intensity from around 0.5 m in diameter up to small tornadoes.
Fire wind	Wind induced by the convective activity of a fire.

Flame angle	The inclination of the flame to the ground surface. Measurement is usually made normal to the fire edge and from the unburnt surface towards the flame.
Flame depth	The width of continuously flaming fuel behind the fire edge and measured normal to the fire edge.
Flame height	The average top height of the flames measured vertically from the ground surface. Unless otherwise specified, the measurement does not include occasional flame flashes which rise above the general level of flames.
Flame length	The length of flames measured along the convective path from the ground surface.
Flank fire	That part of the fire perimeter aligned parallel with the prevailing wind direction.
Fuel moisture content (FMC)	The water content of a fuel particle expressed as a fraction of the oven-dry weight of the fuel particle.
Grassfire	Any fire in which the predominant fuel is grass.
Grassland	Any land on which grasses dominate the vegetation.
Head fire	That portion of a fire edge showing the greatest rate of spread, generally to leeward or upslope.
Heading fire	A fire spreading with the wind, characterised by the flames leaning over unburnt fuels.
Heat transfer	The process by which heat is imparted from one body to another, through conduction, convection or radiation.
Heat yield (low heat of combustion)	The heat of combustion corrected for various heat losses arising mainly from the presence of moisture in the fuel and from incomplete combustion.
Hysteresis	The lag in response to change in causal agent, e.g. the time-lag in response of fuel moisture content to changes in relative humidity.
Indirect attack	A method of suppression in which the control line is located a considerable distance away from the fire's active edge.
Lead tanker	A vehicle equipped with tank, pump and necessary equipment for spraying water and/or chemicals on bushfires, that leads suppression of fire edge in advance of support units.

Litter	The top layer of the forest floor composed of dead loose debris of sticks, branches, twigs, bark and recently fallen leaves or needles, little altered in structure by decomposition.
Mopping-up (mop-up)	Making a fire safe after it has been controlled, by extinguishing or removing burning material along or near the control line.
Oven-dry weight (ODW)	The weight of a fuel sample without adsorbed moisture; usually determined by drying in an oven at 103°C for 24 hours or until the sample reaches a constant weight.
Parts of a fire	On typical free-burning fires the spread is uneven, with the main spread moving with the wind or upslope. The most rapidly moving portion is designated the **head** of the fire, the adjoining portions of the perimeter at right angles to the head are called the **flanks** and the slowest-moving portion is known as the rear or **back**.
Patrol	(1) Generally, to travel over a given route to detect, prevent and suppress fires. (2) More specifically, to go back and forth vigilantly over a length of control line during and/or after construction to prevent breakaways, control spot fires and extinguish overlooked hot spots. (3) Vigilant checking following mop-up until the fireline is considered safe. (4) A person or group of persons who carry out these operations.
Potential rate of spread	The maximum rate of spread that a fire of unlimited width can achieve under the prevailing fuel and weather conditions.
Prohibited period	The period of the year when lighting of fires is controlled under specific legislation.
Psychrometer	An instrument for measuring wet and dry bulb temperature from which the moisture content of the atmosphere can be determined.
Pyrolysis	The thermal or chemical decomposition of fuel at an elevated temperature.
Quasi-steady rate of spread	The mean rate of fire spread, determined by the mean speed of the prevailing wind. If the fire remains narrow the quasi-steady rate of spread will be less than the **potential rate of spread.**

Radiation	(1) The propagation of energy in free space by electromagnetic waves. (2) Transfer of heat through a gas or vacuum other than by heating of the intervening space.
Rate of spread	The relative activity of a fire in extending its horizontal dimensions. Expressed either as rate of increase of the fire perimeter, as a rate of increase in area or as a rate of advance of its head (rate of forward spread), depending on the intended use of the information.
Relative humidity	The fraction (often expressed as a percentage) of the actual vapour pressure of the air to its saturation vapour pressure.
Residence time	Time that flaming combustion persists in the fuel bed; numerically it is the flame depth divided by the rate of spread.
Scrub	A collective term that refers to stands of vegetation dominated by shrubby woody plants or low-growing trees.
Slash	(1) Unusual concentration of fuel resulting from natural events such as wind or snow breakage of trees. (2) Fuels resulting from human activities such as logging, thinning or road construction.
Smouldering fire	A fire burning without flame and barely spreading.
Smoulder time	Time that smouldering combustion persists in the fuel bed.
Spot fire	Fire ignited outside the perimeter of the main fire by flying sparks, embers or larger firebrands.
Spotting	The process of starting new fires beyond the perimeter of the main fire, usually by firebrands that are carried by the wind.
Suppress a fire	Extinguish a fire or confine the area it burns within fixed boundaries.
Threshold wind speed	The wind speed at which a fire moves continuously forward as a heading fire.
Time-lag	See **Hysteresis**.
Veering wind	A wind that changes direction clockwise. Compare with **backing wind**.
Virga	A veil of rain, beneath a cloud, that does not reach the ground.

Wake zone	A zone of light and variable wind downwind of a strong convection column caused by the column obstructing the ambient wind.
Water point	Any natural or artificial source of water that can be used for filling ground tankers.
Weathering	The action of natural atmospheric conditions on any material exposed to them. Weathering includes physical and chemical changes.
Wildfire	See **Bushfire**.
Woodland	Plant communities in which trees, often small and characteristically short-boled relative to their depth of crown, are present but form only an open canopy. The intervening spaces are occupied by a lower vegetation, commonly grass.

Bibliography and further reading

1: Introduction

Beringer J, Packham D and Tapper N (1995) Biomass burning and resulting emissions in the Northern Territory, Australia. *International Journal of Wildland Fire* **5**, 229–235.

Burrows ND, Burbidge AA, Fuller PJ and Behn G (2006) Evidence of altered fire regimes in the Western Desert region of Australia. *Conservation Science W Aust.* **5** (3), 272–284.

Cheney NP (1976) Bushfire disasters in Australia, 1945–1975. *Australian Forestry* **39**, 245–268.

Hallam S (1975) *Fire and Hearth: A Study of Aboriginal Usage and European Usurpation in South-western Australia.* Australian Institute of Aboriginal Studies, Canberra.

Luke RH and McArthur AG (1978) *Bushfires in Australia.* Australian Government Publishing Service, Canberra.

McArthur AG (1960) 'Fire danger rating tables for annual grasslands.' Forestry and Timber Bureau, Commonwealth Department of National Development, Canberra.

McArthur AG (1966) 'Weather and grassland fire behaviour'. Department of National Development, Forestry and Timber Bureau, Canberra. Leaflet 100.

Pyne SJ (1991) *Burning Bush: A Fire History of Australia.* Henry Holt & Co., New York.

Sneeujagt RJ and Peet GB (1985) *Forest Fires Behaviour Tables for Western Australia.* 3rd edn. WA Department of Conservation and Land Management, Perth.

2: Fuel

Bureau of Meteorology (1984) 'Report on the meteorological aspects of the Ash Wednesday fires – 16 February 1983'. Australian Government Publishing Service, Canberra.

Moore RM (1970) *Australian Grasslands.* Australian National University Press, Canberra.

Parrot RT (1964) The growth, senescence and ignitability of annual pastures. MAgSci thesis. University of Adelaide, Department of Agronomy, Waite Institute.

Perry RA (1960) Pasture lands of the Northern Territory, Australia. CSIRO, Melbourne. Land Research Series No. 5.

3: Combustion of grassy fuels

Catchpole EA, de Mestre NJ and Gill AM (1982) Intensity of fire at its perimeter. *Australian Forest Research* **12**, 47–54.

Drysdale D (1985) *An Introduction to Fire Dynamics.* John Wiley & Sons Ltd, Chichester.

Dyer R, Jacklyn P, Partridge I, Russell-Smith J and Williams D (Eds) (2001) *Savanna Burning: Understanding and Using Fire in Northern Australia.* Tropical Savannas CRC, Darwin.

Knight IK and Sullivan AL (2004) A semi-transparent model of bushfire flames to predict radiant heat flux. *International Journal of Wildland Fire* **13**, 201–207.

Sullivan AL, Knight IK and Cheney NP (2002) Predicting the radiant heat flux from burning logs in a forest following a fire. *Australian Forestry* **65**, 59–67.

Sullivan AL, Ellis PF and Knight IK (2003) A review of the use of radiant heat flux models in bushfire applications. *International Journal of Wildland Fire* **12**, 101–110.

4: Fire behaviour

Bradstock RA and Gill AM (1993) Fire in semi-arid, mallee shrublands: size of flames from discrete fuel arrays and their role in the spread of fire. *International Journal of Wildland Fire* **3**, 3–12.

Burrows N, Ward B and Robinson A (1991) Fire behaviour in spinifex fuels on the Gibson Desert Nature Reserve, Western Australia. *Journal of Arid Environments* **20**, 189–204.

Casson NE and Fox JED (1987) The post-fire regeneration responses of *Triodia wiseana* and *T. basedowii. Australian Rangelands Journal* **9**, 53–55.

Cheney NP and Gould JS (1995) Fire growth in grassland fuels. *International Journal of Wildland Fire* **5**, 237–247.

Cheney NP and Gould JS (1997) Letter to the editor: fire growth and acceleration. *International Journal of Wildland Fire* **7**, 1–5.

Cheney NP, Gould JS and Catchpole WR (1993) The influence of fuel, weather and fire shape variables on fire-spread in grasslands. *International Journal of Wildland Fire* **3**, 31–44.

Cheney NP, Gould JS and Catchpole WR (1998) Prediction of fire spread in grasslands. *International Journal of Wildland Fire* **8**, 1–13.

Cheney P, Gould J and McCaw L (2001) The dead-man zone – a neglected area of firefighter safety. *Australian Forestry* **64**, 45–50.

Country Fire Authority, Victoria (1987) *Grassland Curing Guide.* Country Fire Authority, Victoria, Research and Planning Department, Melbourne.

Gill AM, Burrows ND and Bradstock RA (1995) Fire modelling and fire weather in an Australian desert. *CALMScience* Supplement **4**, 29–34.

Griffin GF, Price NF and Portlock HF (1983) Wildfires in the Central Australian rangelands, 1970–1980. *Journal of Environmental Management* **17**, 311–323.

Luke RH and McArthur AG (1978) *Bushfires in Australia*. Australian Government Publishing Service, Canberra.

Marsden-Smedley JB (1993) *Fuel Characteristics and Fire Behaviour in Tasmanian Buttongrass Moorlands*. Parks and Wildlife Service, Department of Environment and Land Management, Hobart.

Marsden-Smedley JB and Catchpole WR (1995) Fire behaviour modelling in Tasmanian buttongrass moorlands II. Fire behaviour. *International Journal of Wildland Fire* **5**, 215–228.

McArthur AG (1967) 'Fire behaviour in eucalypt forests'. Forestry and Timber Bureau, Department of National Development, Leaflet No. 107, Canberra.

McArthur AG (1972) Fire control in the arid and semi-arid lands of Australia. In *The Use of Trees and Shrubs in the Dry Country of Australia*. (Eds N Hall *et al.*) Australian Government Publishing Service, Canberra.

Noble JC (1989) Fire studies in mallee (*Eucalyptus* spp.) communities of western New South Wales: the effects of fires applied in different seasons on herbage productivity and their implications for management. *Australian Journal of Ecology* **14**, 169–187.

Noble JC and Vines RG (1993) Fire studies in mallee (*Eucalyptus* spp.) communities of western New South Wales: grass fuel dynamics and associated weather patterns. *Rangelands Journal* **15**, 270–297.

Noble JC, Tongway DJ, Roper MM and Whitford WG (1996) Fire studies in mallee (*Eucalyptus* spp.) communities of western New South Wales: spatial and temporal fluxes in soil chemistry and soil biology following prescribed fire. *Pacific Conservation Biology* **2**, 398–413.

Saxon EC (Ed.) (1984) *Anticipating the Inevitable: A Patch Burning Strategy for Fire Management at Uluru (Ayers Rock–Mt Olga) National Park*. CSIRO Publishing, Melbourne.

5: Predicting fire spread

Cheney NP and Gould JS (1995) Fire growth in grassland fuels. *International Journal of Wildland Fire* **5**, 237–247.

Cheney NP and Gould JS (1997) Letter to the editor: fire growth and acceleration. *International Journal of Wildland Fire* **7**, 1–5.

Cheney NP and Just TE (1974) 'The behaviour and application of fire in sugar cane in Queensland'. Leaflet 115. Forestry and Timber Bureau, Canberra.

Cheney NP, Gould JS and Catchpole WR (1993) The influence of fuel, weather and fire shape variables on fire-spread in grasslands. *International Journal of Wildland Fire* **3**, 31–44.

Cheney NP, Gould JS and Catchpole WR (1998) Prediction of fire spread in grasslands. *International Journal of Wildland Fire* **8** (1), 1–13.

Coleman J and Sullivan A (1995) 'SiroFire': The CSIRO bushfire spread simulator. In *Proceedings of the Institute of Foresters of Australia 16th Biennial Conference*. 18–21 April 1995, Ballarat, Victoria. pp. 309–319.

Coleman J and Sullivan A (1996) A real-time computer application for the prediction of fire spread across the Australian landscape. *Simulation* **67**, 230–240.

Country Fire Authority, Victoria (1987) *Grassland Curing Guide*. Country Fire Authority, Victoria, Research and Planning Department, Melbourne.

Luke RH and McArthur AG (1978) *Bushfires in Australia*. Australian Government Publishing Service, Canberra.

Paltridge GW and Barber J (1985) Monitoring grassland dryness and fire potential in Australia in NOAA/AVHRR data. *Remote Sensing of the Environment* **25**, 381–394.

6: Local variation and erratic fire behaviour

Bureau of Meteorology (1984) 'Report on the meteorological aspects of the Ash Wednesday fires – 16 February 1983'. Australian Government Publishing Service, Canberra.

Crowder RB (1995) *The Wonders of the Weather*. Australian Government Publishing Service, Canberra.

Marchaj CA (1964) *Sailing Theory and Practice*. Adlard Coles, London.

Sullivan AL and Knight IK (2001) Estimating error in wind speed measurements for experimental fires. *Canadian Journal of Forest Research* **31**, 401–409.

7: Fire danger

Cheney NP and Gould JS (1995) Separating fire spread prediction and fire danger rating. *CALMScience* Supplement **4**, 3–8.

Cheney NP, Wilson AAG and McCaw L (1990) 'Development of an Australian fire danger rating system'. RIRDC Project No. CSF–35A Report (unpublished).

McArthur AG (1966) 'Weather and grassland fire behaviour'. Department of National Development, Forestry and Timber Bureau, Canberra. Leaflet 100.

Purton M (1982) 'Equations for the McArthur Grassland Meter Mk IV'. Bureau of Meteorology. Meteorological Note No. 147.

8: Wildfires and their suppression

Country Fire Authority, Victoria (1983) *The Major Fires Originating 16 February 1983*. Country Fire Authority Victoria, Melbourne.

Gould JS (2006) 'Development of bushfire spread of the Wangary fire, 10th and 11th January 2005, Lower Eyre Peninsula, South Australia'. CSIRO/Ensis, Canberra.

Keeves A and Douglas DR (1983) Forest fires in South Australia on 16 February 1983 and consequent future forest management aims. *Australian Forestry* **46**, 148–162.

Luke RH (1961) *Bush Fire Control in Australia*. Hodder & Stoughton, Melbourne.

Luke RH and McArthur AG (1978) *Bushfires in Australia*. Australian Government Publishing Service, Canberra.

McArthur AG, Cheney NP and Barber J (1982) 'The fires of 12 February 1977 in the Western District of Victoria'. Joint report by CSIRO Division of Forest Research Canberra, ACT, and Country Fire Authority, Melbourne.

Noble JC (1991) Behaviour of a very fast grassland wildfire on the Riverine Plain of south-eastern Australia. *International Journal of Wildland Fire* **1**, 189–196.

Rawson RP, Billing PR and Duncan SF (1983) The 1982–83 forest fires in Victoria. *Australian Forestry* **46**, 163–172.

Wilson AAG (1988) Width of firebreak that is necessary to stop grassfires: some field experiments. *Canadian Journal of Forest Research* **18**, 682–687.

9: Grassfire investigation

Kirk PL and DeHaan JD (1990) *Kirk's Fire Investigation*. Prentice-Hall, Englewood Cliffs, NJ.

10: Myths, facts and fallacies

Cheney NP (1985) Living with fire. In *Think Trees Grow Trees*. Chapter 5. Australian Government Publishing Service, Canberra.

Index

acceleration of fires, *see* fire growth
anchor point, 96, 98
anemogram, 70, 106
anemometer, 42, 58, 71
ash patterns, 108, 110
Ash Wednesday fires, 15, 40, 43, 7, 80, 91, 95, 119, 124
atmospheric
   circulation, 67–72, 78, 88
   instability, 80–82
   inversion, 80
   pressure gradient, 78
   stability, 74, 80, 82
Australian
   fire history, 2–4
   vegetation and fuel types, 7–14

back-burning, 84, 97–100
backing fire, 22–25, 27, 30, 35, 40–43, 60, 72, 76, 108–110, 112, 118
backing rate of spread, 40–43, 60, 72, 76, 112
blocking high, 87
bulk density, 33, 34
burning out, 34, 80, 99, 100, 112
bushfire, 4, 115, 129

carbon dioxide, 18, 19
carbon monoxide, 116
causes of fire, *see* fire ignition
Cavan fire, 101, 102
cellulose, 17–19
charring, 17, 19, 106–108, 110
cloud, 81, 130
cloud cover, 81, 82
cold fronts, 40, 70, 78, 92, 94–97
combustion
   chemistry of, 17–19
   heat of, 21
   incomplete, 19, 23, 26
   triangle, 17, 18
   types of,
      exothermic, 17, 18
      endothermic, 18
      pyrolysis, 19

continuity, *see* fuel
continuous fuel, 2, 4, 5, 8, 14, 35, 49–53, 57, 62, 118
control line, 80, 99
convection, 28, 42, 72, 75, 77, 80, 98, 106, 108, 120–123
convection column, 72, 80, 120–123, 128, 130
curing, *see* grass curing
Cyclone Alby, 43

danger, *see* fire danger
dead fuel moisture content, *see* fuel
determining the origin, 105, 112
dew, 39, 56, 78
differential burning, 108–110
difficulty of suppression, 4, 36, 83–85
   see also suppression
diffusion flames, 18–20, 25, 122
direct attack on fires, *see* suppression
discontinuous fuels, 36, 42, 50, 57
down-draughts, 42, 71, 72, 120

eaten-out pasture condition, 16, 35, 38, 42, 46, 51, 52, 57, 58, 60, 64, 85, 104, 108, 120, 124, 131
electrical pruning, 112, 113
equilibrium moisture content, 56, 78, 79
eyewitness observations, 43, 105, 108, 110–112, 116
Eyre Peninsula fire, *see* Wangary fire

fingers of a fire, 43, 46
fire acceleration, *see* fire growth
fireballs, 127, 128
fire
   behaviour, 5–7, 14, 27, 30–48, 49, 50, 56, 58, 67–82, 105, 112, 116, 121, 129
   ignition, 31, 32, 53, 87, 111–113
   intensity, 4, 7, 21, 29, 30, 38, 99, 103, 108, 122, 131–133
   shape, 6, 22, 35, 36, 73, 106, 108, 112
   rate of spread, 4–6, 27, 29, 31–48, 49, 52–55, 57–65, 72, 76, 85, 128, 129
   effect of slope, *see* slope

firebreaks, 4, 57, 58, 64, 87, 96, 102–104
    construction, 102, 103
    effectiveness, 103, 104
fire causes, *see* fire ignition
fire damage, 4, 87, 92, 116, 130,131, 133
fire danger, 4, 15, 56–58, 83–85, 88, 90, 95, 97, 100, 101, 104, 112, 113, 131
    index, 5, 83–85, 89, 95, 132
    rating, 4, 83, 84
    rating systems, 4, 83–85
    meter, CSIRO-modified McArthur Mk 4, 84
    meter, McArthur Mk 4, 5, 56
    meter, McArthur Mk 5, 5
fire growth, 32–35
firefighters, 5, 6, 27, 29, 34, 56, 58, 67, 71, 77–81, 90, 91–100, 112, 127
firefighting, *see* suppression
fire ignition
    spontaneous, 32
    power line, 31, 112, 113
    glass bottles, 32
    metal on rock, 31
    see also fire behaviour
fire intensity, 4, 7, 14, 21, 27, 29, 30, 38, 99, 103, 108, 110, 115, 121, 122, 127, 131–133
fire protection, 115–120, 124–127, 130, 131
fire shape, 6, 12, 22, 35, 36, 73, 106, 108–112
fire spread, 31, 35–47, 49–65, 104, 108–112
    meter, 62–64
    meter for Northern Australia, 64–65
    rate of, 5, 6, 31, 35–48, 49–65, 72, 76, 85, 90, 93–96
    prediction of, 5, 6, 31, 42, 45, 49–65, 67, 85
fire weather, 4–6, 22, 26, 58, 80, 85, 87, 90, 92, 95, 96, 100, 120, 131
fire whirlwinds, 80, 81, 120, 122
fires, types of, 21–25
    head, 4, 12, 22–25, 27–30, 32–35, 41, 43, 50, 63, 72, 73, 84, 87, 96–100, 102–104, 106, 111, 116–118, 121
    back, 22–25, 27, 29, 30, 40–42, 108–110, 112, 118
    flank, 2, 13, 22–25, 27, 29, 34, 43, 47, 96–104, 108–112, 118, 129
flame dimensions
    depth, 25–27, 38, 83, 118
    height, 6, 7, 21, 24–26, 36–38, 57, 83, 110, 124, 128
    length, 24–26, 29, 44, 45

flame front, 20, 27–29, 31, 35, 42, 72, 74, 97, 115, 120, 121, 123, 125–129
flanking fires, 22, 24, 25, 27, 29, 34, 35, 43, 96, 98, 99, 112, 118
forest fires, 1, 16, 30, 80, 81, 118, 121, 127
fronts, *see* cold fronts
fuel,
    amount, 8, 10, 14, 29, 47, 55, 85, 120, 128, 131, 133
    characteristics, 5, 7, 35–38, 47, 83
    condition, 35–38, 49–52, 57, 85
    continuity, 7, 14, 35–37, 40, 45, 58
    curing, *see* grass curing
    distribution, 36, 44, 45, 128
    litter, 8, 22, 65, 80, 125
    load, 7, 12, 27, 37, 38, 45, 48, 85, 120, 125–126
    management, 44,
    moisture content, 38–41
        dead fuel, 21, 31, 39–41, 44–46, 49, 55, 56, 72–74, 78, 79
        equilibrium, 56, 78, 79
        live, 38, 39, 45, *see also* grass curing
        saturation, 78, 79
    types, 1, 7–14, 27, 30, 50–52
    weight, *see* fuel load

grass curing, 38, 39, 49, 52–55
    function, 39, 55, 72–74, 123
    across the landscape, 52–55
    from satellite images, 54, 55
    state, 49, 52–55
grass types
    tropical, 7–9, 41
    tussock, 9, 10, 28, 36, 41, 55, 95, 101, 133
    hummock, 10–12, 35, 36, 39, 44–48
    improved pastures, 1, 12–14, 50
    crop lands, 1, 13, 14, 28, 52,
grassland life cycles, 14, 15
grazed pasture condition, 14, 36, 38, 42, 50, 51, 57, 62–64

head fire, 4, 12, 27, 29, 32–35, 43, 63, 72–74, 96–100, 103, 104, 106, 111, 116–118, 121
    parabolic, 73
    pointed, 72
heading fires, 22–25, 40–43, 46, 108–112
heat yield, 21, 29, 129, 130
heat transfer,
    conduction, 28
    convection, 26, 28, 42, 68, 77, 97, 120–123

radiation, 20, 27–29, 112, 115–119, 126
humidity, *see* relative humidity
hummock grasslands, 10–12, 35–40, 44–48
hysteresis loop, 78–80

ignition sources, *see* fire ignition
improved pastures, 1, 12–14, 50
indirect attack, 99, 100
  see also suppression
initial attack, 96
  see also suppression
instability, 80, 81
investigation of fires, 105–114

katabatic wind, 77

land breezes, 77–78
leaf freeze, 108
leaf scorch, 108–110
live fuel moisture content, *see* fuel moisture content

Mangoplah fire, 88–90
McArthur Fire Danger Rating Systems, 4–6, 56, 83–85
McArthur Grassland Fire Danger Meter, 5, 56
moisture content, *see* fuel
mopping up, 2, 12, 96, 98, 100–104
mown grass, *see* grazed pasture condition

Narraweena fire, 5, 91–95, 100, 109, 124, 128
natural pasture condition, 21, 36–38, 50–52, 57, 60, 61, 64, 65, 72–74

observations, eyewitness, 43, 105, 108, 110–112, 116
open forests, 1, 2, 7–9, 27, 33, 50, 52, 58–60, 64, 65, 75, 109
oven-dry weight, 21, 39, 55,

parts of a fire, 22
petrol tanks, chances of exploding, 124–125
pine plantations, 4, 93, 109, 122, 132
power lines, 31, 90, 111–113,
precipitation, *see* rainfall
predicting fire spread, 4–7, 31, 42, 45, 49–65, 67, 85,
prescribed burning, 80, 118,
Pura Pura fire, 101
psychrometer, 56

radiation, *see* heat transfer
rainfall, 2, 4, 7, 14–16, 45, 47, 48, 50, 53, 56
  deficiency, 15, 16
rate of spread, 4–6, 27, 29, 31–48, 49, 52–55, 57–65, 72, 76, 85, 128, 129
  backing, 23, 24, 35, 40–42, 60, 72, 76, 109
  forward, 32, 34, 38–44, 47, 49, 57–65, 85
relative humidity, 5, 39, 45, 55, 56, 78–81, 83
reaction zone, 19, 20, 122, 123
residence time, 26–28, 38,

safety and survival
  in houses, 117, 127, 131
  in the open, 27, 95, 117–119
  in vehicles, 124–127
scorch height, 109, 110
sea breeze, 34, 77, 78, 97
shrubs, 65, 108–110, 118
slope, 43, 44, 49, 59–62, 74–77, 81, 121–123
smoulder time, 10, 27, 28
solar radiation, 50, 81, 82,
spinifex
  rate of spread in, 44–48
  effect of rainfall, 47, 48
  discontinuity, 11, 12, 44, 45
  fuel distribution, 10–12, 45
  fuel moisture content, 39, 40, 45, 46
  hummock grasslands, 10–12, 35–40, 44–48
  threshold wind speed, 42, 46, 47
spontaneous combustion, *see* fire ignition
spotting, 76, 80, 81, 99, 100, 102
stability, 74, 80, 82
stem freeze, 108
Stevenson screen, 56
stubble, 13, 14, 34, 52
suppression
  difficulty of, 4, 7, 34, 36, 76, 83–85,
  fast moving fires, 96–100
  methods of attack
    direct, 84, 96, 97
    extended, 96, 97
    flank, 98, 99
    head fire, 97, 98
    indirect, 99, 100
    initial, 96
  mop-up, 2, 12, 96, 98, 100–104
  patrol, 100, 101
  slow-moving fires, 97
  tankers, 96, 98
survival in fires, *see* safety and survival

synoptic situations, 77, 87, 88

tankers, *see* suppression
Tatyoon-Streatham fire, 90–93
temperature,
    air, 45, 55, 56, 78–80, 83, 97, 117–119
    combustion, 17–19
    flame, 19–21, 117, 122, 123
terrain, 44, 59–62, 74–77
thermal,
    flames, 17–20, 117, 118, 123
    convective, 27, 42, 71–74, 77, 121
thermometer, 56
threshold wind speed
    continuous fuels, 41, 42, 57, 58
    discontinuous fuels, 11, 36, 46–48
thunderstorm down-draughts, 81, 82
time-lag, 78, 79
tree-lined roads, 90, 99, 102, 103, 125, 127
tropical grasslands, 7–9, 41
tussock grasslands, 9, 10, 28, 36, 41, 55, 95, 101, 133

undisturbed pasture condition, 50, 57, 65, 73, 74
ungrazed pasture condition, 9, 14, 36, 38, 40, 50–52, 58, 60, 64, 65
up-draughts, 42, 68–75

virga, 81
visibility, 81, 111, 116, 119, 128

Wallinduc-Cressy fire, 90, 91
Wangary fire, 95
weather, 3–5, 58, 67–82, 83–85, 87, 88, 120–123
weather observation standards, 56
Western District fires, 90–93
wheat stubble, 13, 14, 34, 52

whirls and whirlwinds, 80, 81, 120, 122
wildfires, 1, 30, 31, 34, 42, 52, 58, 63, 87–104, 112, 122, 131
wind
    anabatic, 77
    average or mean, 32–35, 40–42, 46, 47, 57–60, 67–72, 74, 92,
    conversion from 2 m to 10 m, 57–60
    change in, 5, 24, 27, 33, 34, 40, 42, 43, 47, 49, 67–78, 81, 82, 87, 90–97, 101, 106, 109, 118, 119, 129
    direction, 5, 22, 24, 27, 29, 32–35, 42, 44, 46, 47, 58, 60, 67–78, 81, 82, 87, 88, 96, 99, 104, 106–109, 118, 129, 131
    down-draughts, 42, 71–74
    eddies, 42, 44, 60, 67–72, 76, 77, 106, 107, 121,
    field, 72, 74, 76, 120
    fluctuations, 22, 24, 26, 29, 32, 34, 40, 46, 108, 109
    gusts, 24, 40, 43, 49, 57, 68–72, 92, 121, 129
    katabatic, 77
    local winds, 22, 58, 67–71, 77, 106, 108
    lulls, 27, 49, 57, 68–72, 77, 96, 108, 110, 119, 121
    measurement, 40, 57, 71
    recurved, 74, 76
    speed, 5, 11, 27, 32–37, 40–47, 49, 52, 55, 57–63, 67–78, 83, 85, 89, 95, 102, 110, 117, 120–124
    speed threshold, *see* threshold wind speed
    speed variation, 27, 49, 63, 67–71, 74
    streamwise, 44, 60, 74, 76
    terrain, 44, 74–77
    turbulence, 20, 26, 67–71, 74, 81, 123
woodlands, 1, 2, 7–9, 12, 25, 27, 50, 52, 58–60, 64, 65, 73